The Ecology Action Guide

Graham C. Hickman
&
Susan M. Hickman

Benjamin
Cummings

San Francisco • Boston • New York • Cape Town • Hong Kong
London • Madrid • Mexico City • Montreal • Munich
Paris • Singapore • Syndey • Tokyo • Toronto

Acquisitions Editor: Elizabeth Fogarty
Publishing Assistant: Jeanne Zalesky
Marketing Manager: Josh Frost
Design and Production: Joan Keyes, Dovetail Publishing Services
Cartoons: Josh Frost
Manufacturing Buyer: Vivian McDougal
Cover Designer: Andrew Ogus
Cover photo: Graham C. Hickman

ISBN 0-321-06883-1

8 9 10 CRS 07 06 05

Contents

Preface

It is important to keep a positive attitude. Sure, we can be apprehensive, and maybe even angry, over the state of some of our natural resources, but each and every one of us has the opportunity to do something positive, something to make things right, and even a little better than before. By channeling our energy in a constructive way, we can all enjoy a more interesting and healthy planet, and life.

This Ecology Action Guide is written for students, concerned citizens, politicians, and anyone compelled to do something to help make the world a better place to live. Effective action may involve emotional appeal but must also employ fact-based, quantitative data. By understanding basic principles of ecology, we can be more successful in applying strategies to the betterment of our environment. In understanding how ecosystems function, we have more motivation to keep ecosystems working normally instead of blindly following lists of things to do. Knowledge and direction are keys to finding solutions to problems.

Because this is only a guide, coverage of many topics is brief. The introductory chapter vignettes are meant to spark discussion and set the scene for the references that follow, but are far from providing an exhaustive treatment of the topics. The inquiring mind will need to seek more details in the many basic ecology textbooks, a listing of which is given in Chapter 7. On the other hand, some nonscience readers may find that we are including too much basic material for a sophomore or junior course. We have tried to stay on the middle path, but one book cannot cover all the needs of ecology and environmental biology classes from freshman through graduate levels, both for science and non-science majors. The references, however, will apply to all levels of study depending on how the reader wishes to employ them.

Beyond theoretical considerations, this guide is a resource for practical information not found in many textbooks, such as sources of equipment, media companies, websites, and other contacts to individuals and organizations with interests similar to your own.

Only general suggestions are provided to finding solutions. What you can accomplish and how you accomplish it is your responsibility. Always keep in mind that civility and a regard for another point of view are the basis for reaching workable solutions.

The authors and publisher cannot be held responsible for the service, quality, or availability of materials mentioned in this guide. It is a free market, and you should seek the company or vendor (not necessarily in this guide) that best serves your individual purposes and requirements. There has been no payment or pressure from any political or entrepreneurial organization for inclusion in this book. Listings are based on the experience and preferences lof the authors. We would appreciate your suggestions for making the listings more complete and useful. Likewise, we would like to know which sections or references were not as useful and which you would exclude.

Any errors in references, phone numbers, and postal and web addresses are very much regretted. We also realize that contact numbers and addresses will change from time to time. The short-term solution has been to provide redundancy in contact information with addresses, telephone numbers, website addresses, and e-mail addresses. In addition, Benjamin/Cummings will update the reference lists on-line at http://www.awl.com every year. Please contact the authors at the address below with any corrections, updates, or suggestions for future editions. We very much appreciate your input.

In keeping with the general tone of this book, please select your own motivational quotation from one of the lists within, or try:

Rodes, Barbara K. and Rice Odell, compilers. 1992. *A Dictionary of Environmental Quotations*. Simon & Schuster: New York, NY.

Graham and Susan Hickman
Department of Physical and Life Sciences
Texas A&M University–Corpus Christi
6300 Ocean Drive
Corpus Christi, TX 78412
ghickman@falcon.tamucc.edu

Acknowledgments

Our special thanks to Jeanne Zalesky of Benjamin/Cummings Publishing Company for drawing this project to our attention, for providing support and guidance, and for being a pleasure to work with. Ann Causey's action guide was the seed from which this project grew, and we appreciate her many ideas and thoughts on many topics. Referees offered their corrections, comments, opinions, information, and insight, all of which are very much appreciated. John and Lara Hickman, college student seekers of the truth, saved us much time and consternation by updating many of the references and URLs. John Snyder's *Nature Photography* (jrsnywlp@infi.net) provided valuable references and advice, as well as motivation to make our photographs as professional as possible. Special thanks to the webmasters who had all of the contact information for their organizations in one easily accessible place. Also thanks to the webmasters who may be reading this and are going to group all of their contact information in one area of their website. Lastly, thanks to the students in our classes in Ecology, Environmental Science, Physiological Ecology, Padre Island Ecology, and Vertebrate Biology for providing new enthusiasm and a fresh perspective every semester.

Chapter 1

Introduction

One person *can* make a difference in the outcome of an event without having a lot of money or sacrificing that which each of us considers important in our own individual way. That is not to say that all will be made well in the world at once, but instead, that progress will be made toward improving the situation.

What is vital to success in almost any endeavor is knowledge, time, persistence, and heart (motivation). This guide is meant to encourage learning about ecosystems and ecosystem functioning, to suggest some simple things that can be done to keep ecosystems healthy, and to give an understanding as to how and why these methods and practices of ecosystem care are effective. To this end, extensive suggestions are provided for becoming more involved in ecological issues.

What Is Ecology?

Many disagreements and misunderstandings are the result of semantics (the definitions given to words and the change in the meaning of words with the passage of time). The term *ecology* is one term in particular that has been broadly applied by both scientists and nonscientists. If students in a class are asked the definition of *ecology*, many different answers are likely to result. There is also a large divergence of meaning of the term *ecology* in textbooks, with various authors emphasizing ecosystems, species interaction, populations, distributions, and so on.

Also, terms become modified over ltime so that a definition must be viewed in terms of the period of time in which the term is being used. The definition of ecology given by Haeckel in the 1800s is outdated in terms of our current understanding of ecology. Terms may then acquire several meanings. A bald eagle, for example, is not lacking in head covering; bald in this instance simply means white.

Terms are also "flavored" by different perspectives because every person has a different way of looking at and interpreting things based on past experiences and training. For example, a group of scientists

Table 1.1   *Important Ecological Journals and Their Inaugural Year*
                (Note that ecology is a relatively young discipline.)

General Ecological Journals

Advances in Ecological Research	1962	Academic Press
Annual Review of Ecology and Systematics	1984	Annual Reviews
Ecological Monographs	1930	Ecological Society of America
Ecology	1920	Ecological Society of America
Journal of Ecology	1912	British Ecological Society
Oecologia	1963	International Association for Ecology
Oikos	1949	The Nordic Society, Oikos

Specialty Journals in Ecology

Behavioral Ecology and Sociobiology	1976	Springer-Verlag, NY
Ecography	1992	Munksgaard
Ecological Applications	1990	Ecological Society of America
Ecological Economics	1989	Internat. Society for Ecological Economics
Ecological Entomology	1976	Royal Entomological Soc. London
Ecological Modeling	1975	Elsevier, Amsterdam, Netherlands
Ecology Law Quarterly	1971	School of Law, University of California, Berkeley
Evolutionary Ecology	1989	Evolutionary Ecology, Ltd.
Evolutionary Ecology Research	1999	Evolutionary Ecology, Ltd.
Human Ecology	1972	Plenum, NY
Journal of Animal Ecology	1933	British Ecol. Society, Blackwell, Oxford

riding down the highway in a car may view the countryside in very different ways: a botanist would be looking at the greenery, a geologist at the rock formations in the road cut, a zoologist at birds, a meteorologist at the clouds, and so forth. The list of ecological journals in Table 1.1 demonstrates the wide diversity and interests of ecologists studying natural communities in order to determine how the natural world works. Both ecology and environmental science have similarities with economics, which is the study of the production and consumption of resources in a society.

Any ecological terms used in this guide that you do not know or that are being used in a way that is not familiar can be found in the glossary of one of the many ecology textbooks listed in Chapter 7.

For the sake of simplicity, **ecology** will be defined in this book as the study of the

Table 1.1 *(continued)*

Journal of Applied Ecology	1966	British Ecol. Society, Blackwell, Oxford
Journal of Chemical Ecology	1975	Springer-Verlag, NY
Microbial Ecology	1975	Springer-Verlag, NY
Restoration Ecology	1993	Blackwell Scientific Publishers
Restoration and Management Notes	1981	University of Wisconsin Press
Urban Ecology	1975	Elsevier, NY
Web Ecology	2000	Munksgaard

Specialty Journals in Various Geographical Areas and Habitats

Australian Journal of Ecology	1976	Australian Ecological Society, Blackwell Sci. Pub.
Holarctic Ecology	1978	Munksgaard, Copenhagen
Journal of East African Wildlife Ecology	1985	Ecological Society of East Africa
Journal of Tropical Ecology	1985	International Association of Ecology
Acta Oecologica Scandinavica	1950	Munksgaard, Copenhagen
Soviet Journal of Ecology	1970	Consultants Bureau, NY

Aquatic and Marine Ecology Journals

Journal of Exp. Marine Biology and Ecology	1963	North Holland
Journal of Freshwater Ecology	1981	Oikos Publishers
Limnology and Oceanography	1956	Am. Society of Limnology & Oceanography
Marine Ecology	1969	Paul Parier, Napoli, Italy

interactions between living things and their environment. Topics include many facets of study. *Population regulation*, for example, can be affected by competition, predation, disease, and parasites. *Species richness* involves why and how species are distributed. *Energy flow* entails the production of foods from solar energy and the distribution of this energy along with nutrients through the environment. *Communities of interacting species* and *populations of individuals* are other major topics of study.

Environmental science more specifically deals with the impact of humans on these natural systems, including the principles of ecology and the development of sustainable solutions to environmental problems. Topics include agronomy, soils, agriculture, aquaculture and fisheries, hydrology, energy, human populations, and legal compliance, among many others.

Ecology has been seen as the basic science. Environmental science may use ecological principles and may also involve statistics, meteorology, soils, the flow of materials through an ecosystem, and many other topics. Environmental science may also include legalities, ethics, sociology, and other areas focusing on humans.

It is important to realize that the main objective in using terms is the communication of ideas and that techniques and ideas of one discipline may well be applicable to other disciplines. Life table techniques developed for insurance purposes in humans are useful for the study of animal populations, and the concept of carrying capacity developed in ecology can be used in the study of human populations. Sometimes too much compartmentalization can be a barrier to innovative thinking. Both ecology and environmental science are multidisciplinary, with strong input from chemistry, geology, mathematics, biology, and related sciences.

A strong environmental science influence is found throughout the guide. Many students find it motivating to see how the knowledge they are acquiring applies to everyday living. Also, many of the job opportunities in ecological fields relate strongly to environmental science. Graduates in the ecological field are assuming important positions in industry and government. It is important that ecology students be familiar with current issues, look at both sides of an argument, be able to procure the pertinent data, and make rational decisions based on facts rather than relying purely on emotion.

For further ideas on defining ecology, see:

Owen, Denis Frank. 1974. *What Is Ecology?* Oxford University Press: New York, NY.

Science in the Real World

If ecology as a basic science were simply rational and logical thinking, it would be a much easier science to study. However, ecologists also have a responsibility to provide sound science in moral and ethical decisions (Should millions of dollars be spent to save an endangered species while people are starving?), economic questions (Should a nation's economy suffer through the conservation of tropical rain forests?), and political (Will it get me votes?) issues.

Unfortunately, the majority of public officials who have the responsibility for making these important decisions have had little or no training in ecological science. This is analogous to taking your sick dog to a mechanic instead of a veterinarian. We need involved and caring scientists and an educated and informed citizenry to help our elected officials make the proper decisions on important ecological questions. We also need to look beyond our own horizons at global patterns. Ecological processes do not stop at political boundaries.

Formal Study and Employment

Before taking any courses, it is advisable to speak with a curriculum advisor to see how the courses you want to take fit in with your goals and the programs offered at your school. Because ecology is a composite science, some of the science foundation courses, such as mathematics (especially

technicians. The School for Field Studies and the Student Conservation Corps enable students to obtain college credit during the summer. The Organization of Biological Field Stations (OBFS) consists of approximately 180 field stations and professionals concerned with field studies primarily in North and Central America. Most OBFS stations offer course credit for study. The experience you gain will be invaluable when you begin your job search. Search by location or other criteria at:

http://www.obfs.org *and* obfs@ucdavis.edu

The School for Field Studies (SFS) offers undergraduate courses in environmental studies, conservation biology, and resource management with programs located around the world (including Australia, Kenya, and Costa Rica). Hands-on fieldwork allows students to appreciate the interaction of science, social factors, and economics in environmental studies. More information can be requested from:

The School for Field Studies, 16 Broadway, Beverly, MA 01915
   800-989-4435
   http://www.fieldstudies.org *and*
   admissions@fieldstudies.org

More ecologically oriented jobs are available today as a result of the expanding human population and increased regulations. Positions might involve outdoor recreation, consulting, education, national park guides, wildlife management, animal behavior, horticulture and agriculture,

statistics), chemistry, biology, geology, and computer science, will be helpful. Courses in ecology, environmental science, microbiology, and field techniques are fundamental.

Graduate programs in ecology vary among universities, so browsing graduate school catalogs is recommended. Pertinent courses might include GIS (Geographic Information Systems) computing, regulations and laws, meteorology, toxicology, conservation, restoration, wetlands, marine biology, animal behavior, evolution, and perhaps some undergraduate courses. An academic advisor can assist you in choosing suitable courses for your goals.

It is not always easy to decide what you would like to do for the rest of your life. Undertaking volunteer work and internships at national parks, aquaria, zoological parks, industries, and public offices can help you focus on areas of interest, as well as eliminate areas that you thought looked great on paper.

Summer jobs are available at state agricultural research stations as lab assistants or urban ecology, marine ecology, teaching, and local and national work in research, regulations, issuing permits, and testing. Specializations are also available within diverse fields such as environmental journalism and environmental law. Table 1.2 on pages 8 and 9 gives a more complete listing of career opportunities.

Because these fields are so diverse, the amount of financial compensation for positions varies greatly. Also, within different types of positions, there is a broad degree of variation, depending on the amount of education and training required, the experience, the area of the country, the size of the institution, and so forth. If financial considerations are a top priority, it is important that you check the pay scale before you begin your program of study. As a rough rule, the more desirable the job, the more competition there will be. With more applications may come a decrease in pay (the law of supply and demand). It is advisable to have a number of qualifications. Attain a degree (preferably graduate school) and gain as much skill and work-related experience as possible through internships, volunteer work, and related employment positions.

Careers in Ecology brochures are available from the Ecological Society of America:

High school
    http://esa.sdsc.edu/highschool.htm

Undergraduate
    http://esa.sdsc.edu/undergraduate.html

Some Career Profiles

The following brief accounts will answer potential questions about decisions and directions in pursuing ecological studies.

Damon Williford, Senior Biology Undergraduate

I first developed an interest in animals and biology in general when I was four after my first trip to a museum. I was unsure of what I wanted to do after graduating from high school, so I entered the Air Force, and served two years on active duty and four in the Reserves. By 1997 I knew what I wanted to do and enrolled first at Del Mar College and then transferred to Texas A&M University–Corpus Christi.

Currently, I work at the Texas State Aquarium as an animal husbandry assistant, a job that stemmed from an internship. My responsibilities include the feeding and care of exhibit animals (marine fishes, alligators, raptors, shorebirds, and otters) and the maintenance of their exhibits, as well as performing daily educational programs about them aimed at raising the public's awareness about conservation and ecology. Additional responsibilities include the care of injured and sick wild birds that are undergoing rehabilitation.

Next August I will begin field research for my thesis aimed at learning more about the winter ecology of burrowing owls in South Texas. This research has several goals: to determine the diet of the owls by collecting and examining owl pellets, to determine whether the owls exhibit site fidelity and identify what factors influence burrow selection, and to gather data on the weight of the owls throughout their winter stay. Determining site fidelity and weighing the birds will involve trapping and banding them so that the same individuals can be tracked over the winter.

My advice to students interested in a career in biology is to get all the hands-on experience they can. Knowledge gained in the classroom is important, and having a 4.0 GPA is impressive. However, without the field experience one has no idea of how everything that was learned in class relates to the world outside. It is one thing to read about how to band a bird or set up an exhibit, and quite another to actually do it!

The best part of ecology and field biology is that I can be outside. I shudder at the thought of having to spend my whole life behind a desk in a suit and tie!

Ellen Swepston, Biology Graduate Student

My initial interest in science began in junior high school and intensified throughout high school. During college, the diversity of ecological studies became my primary interest. Surrounded by fellow students who focused on only one aspect of biology, I realized that I wanted to see the "whole picture." The college classes involving fieldwork were a real help in understanding the broader scope of ecology. Seeing how everything intertwines is the part I enjoy most in biology.

After graduating with a B.S. in biology, I realized that I would need an advanced degree to successfully make my mark. However, I was somewhat burned out on academics. Instead of immediately entering graduate school, I chose to take a year off, move home to Houston, and work. It took a mere two months of working in a professional capacity to realize that I was not as burned out on academics as I had previously thought. Working in an office 40 hours a week quickly had me reevaluating my life and career goals. I enjoyed my work and I was good at it, but it did not satisfy my thirst for science.

I returned to graduate school at Texas A&M University–Corpus Christi to experience the numerous field opportunities available. I am working on an M.S. in biology and an M.S. in secondary education so that I can spread my enthusiasm to future ecologists.

During my career as a graduate student, I have been fortunate to conduct and assist in numerous field studies. Within months of entering the graduate program, a professor put me in touch with the United States Geological Survey (USGS) office housed on campus. I assisted with the research that the USGS was conducting at Atascosa Wildlife Refuge in South Texas, which involved bird counts, collection, identification, and recording, as well as vegetative surveys. It was an incredible experience that allowed me to grow as an ecologist. The course work has also allowed me to study the local ecology of the Texas coast and the ecology of Mexico, both coastal and inland. One research project involved monitoring the ecology of a reef system using nondestructive photographic methods.

I am currently researching a project involving the collection, identification, determination of distribution patterns, and propagation of seeds from corridor transects on South Padre Island National Seashore. Upon graduation, I plan to enter the workforce as a teacher with the goal of interesting students in ecological studies.

Table 1.2   *Career Opportunities in Ecology*
                 (Note that the professions listed may apply to more than one category, and that
                 this list is not comprehensive but only suggestive of possible avenues.)

General Ecological Vocations

Biostatistician	Statistics of ecosystems; populations; predator/prey, host/pathogen, plant/herbivore interactions
Botanist	Plant identification, community structure, resource dynamics, genetic engineering, new plant varieties
Ecologist	Understanding of how natural and impacted systems function
Educator	Train and inform students and citizenry
Landscape architect	Designing parks and other facilities that preserve or restore ecosystems
Microbiologist	Bacteria, viruses, other microorganisms in food chains, bioremediation, diseases
Zoologist	Determining the requirements of animal species and their part in ecosystems, communities, populations

Vocations Focusing on General Environmental Science

Chemical engineer	Design of technology for disposal of chemical wastes
Chemist	Detection and identification of pollutants
Demographer	Application of statistics to populations
Environmental health scientist	Determining ecosystem health
Environmental health specialist	Review EPA regulation compliance, impacts on human health
Epidemiologist	Study of the spread of disease in populations
Hazardous waste manager	Proper disposal of hazardous wastes
Industrial hygienist	Responsible for healthy work conditions
Risk manager	Evaluating the risk to the environment posed by industrial activities for insurance purposes
Solid waste manager	Overseeing the transport and disposal of solid wastes
Toxicologist	Tracing the link between illnesses and toxins in the environment
Water quality technologist	Contaminant testing at water treatment plants

Vocations Focusing on the Abiotic (Nonliving) Sector

Air quality engineer	Control of air pollution
Atmosphere scientist	Determining the presence of chemicals in the atmosphere
Geological engineer	Designing landfills and disposal sites for hazardous waste

Table 1.2 *(continued)*

Vocations Focusing on the Abiotic (Nonliving) Sector (continued)

Hydrologist	Studying water resources
Meteorologist	Study of the atmosphere and weather
Mining engineer	Minimizing methane and groundwater leaks
Oceanographer	Testing for the extent and distribution of pollutants
Petroleum engineer	Minimizing the impact of drilling; restoration of drilling sites
Soil scientist	Monitoring landfill sites

Vocations Focusing on the Biotic (Living) Sector

Agricultural engineer	Designing technology to conserve soil and water
Agronomist	Research alternatives to chemical fertilizers and pesticides
Aquaculture	Raising fish and aquatic/marine invertebrates such as shrimp for the table
Fisheries biologist/manager	Growth and survival studies on fishes
Forestry/park ranger	Tree planting and maintenance, education, resource management
Interpretive naturalist	Lead educational tours in parks, zoos, and natural areas
Wildlife biologist/manager	Preservation and management of wildlife populations and habitats

Vocations for the General Sector

Architect	Design energy efficient buildings
Civil engineer	Design public works projects involving highway and sewage restoration
Computer specialist	Monitoring systems and computer simulations
Electrical engineer	Conservation of electrical energy
Emergency response specialist	Chemical and oil cleanups
Environmental lawyer	Interpretation of laws, regulations, enforcement
Journalist	Writing on scientific issues
Mechanical engineer	Developing environmental-compliant technology
Nuclear engineer	Nuclear power plant design and disposal of nuclear waste
Occupational physician	Radioactive and toxin exposure treatment
Technical writer	Creation of environmental impact reports
Urban community planner	Planning urban areas to minimize ecosystem disruption

Dr. Wes Tunnell, Professor of Biology and Director of the Center for Coastal Studies, Texas A&M University–Corpus Christi

I seem to have been born with a love of the outdoors. Growing up in South Texas, I spent much time on farms and ranches, as well as bays and beaches, with my father. I developed biological interests, particularly field studies, while earning a bachelor's degree at Texas A&M University–Kingsville.

After graduation, I stayed in Kingsville to work on a master's degree involving marine fieldwork, research, and travel in Mexico. My M.S. thesis focused on the molluscan (seashell) ecology of a reef offshore of the newly established Padre Island National Seashore during 1968–1969. My early publications involved a broad interest in the natural sciences and a sense of discovery, including vertebrate Pleistocene fossils from the reef study site, molluscan ecology and distribution, brachiopod (lamp shell) taxonomy, and reef formation theory.

After completing an M.S. in 1969, I was drafted into the U.S. Army where I served two years as a biological laboratory assistant in the Medical Corps at Ft. Baker, Sausalito, California. Taking full advantage of this new geographic locality, I studied at the University of California–Berkeley at night. I traveled the entire West Coast with my young family from Baja Mexico to Oregon, diving and collecting molluscs. After military duty, I moved across country to Tampa, Florida, and the University of South Florida to begin a Ph.D.

Upon arriving in Florida, a very significant occurrence happened that would influence and direct the rest of my career.

Dr. Donald R. Moore, an international micromollusc specialist from the University of Miami's Rosentiel School of Marine and Atmospheric Sciences, whom I had corresponded with during my M.S. work, invited me on a scientific cruise to the Bahamas to study molluscs. This two-week cruise of all day/every day, shallow-water snorkeling introduced me to tropical marine biology. This theme of introducing students to the tropics was later to become a career goal in my teaching. After one year in Florida, I returned to Texas A&M University where I continued my Ph.D. in biology. My dissertation research involved molluscan systematics, distribution, and ecology of two coral reefs offshore of Vera Cruz, Mexico, in the southwestern Gulf of Mexico.

Upon graduation with a Ph.D. in 1974, I started an academic career as an assistant professor of biology at Texas A&M University–Corpus Christi. I taught many classes, such as Marine Biology, Marine Ecology, Ichthyology, Estuarine Biology, and Texas Coastal Ecology. However, my most famous class is Coral Reef Ecology. Now in its 26th year, it includes a two-week field trip to Mexico. Moving from assistant to associate and to full professor, I have conducted research all along the Texas coast, as well as in Mexico. In 1984 I founded the Center for Coastal Studies, a marine science research institute at Texas A&M University–Corpus Christi.

I advise students to develop a passion for some aspect of study and become a real expert in that area. At the same time, however, I recommend a broad base of studies and understanding, because you never know what you will eventually be doing.

Developing a love for science and having broad interests will keep one's mind and energy active for years, seeking and discovering new aspects of the natural world.

More information on what an ecologist does and career profiles of some practicing ecologists can be found at:

http://wfscnet.tamu.edu/jboard/careersearch.htm

Career and Job Opportunities Reference List

Additional contacts can be found on the websites of the many agencies and organizations that are listed in this guide. Jobs with the federal government may have very narrow windows for applying for positions.

American Fisheries Society Jobs Page
    http://www.fisheries.org/
    employ_job_links.htm *and*
    jobs@fisheries.org

Careerbuilder (interview, posting your resume, e-mail job listings)
    http://www.careerbuilder.com/

Careers in Natural Resources and Environmental Law. Luney, Percy. 1987. American Bar Association: Chicago, IL.

Chronicle of Higher Education, 1255 23rd St. NW, Suite 700, Washington, DC 20037
    202-466-1000
    http://thisweek.chronicle.com *and*
    help@chronicle.com

The Complete Guide to Environmental Careers in the 21st Century. Environmental Careers Organization. 1999. Island Press: Covela, CA.

Earth Work Magazine, Student Conservation Association, P.O. Box 550, Charlestown, NH 03603
    603-543-1700
    http://www.sca-inc.org/ *and*
    earthwork@sca-inc.org

Ecological Society of America, 1707 H St. NW, Suite 400, Washington, DC 20006
    202-833-8773 or Public Affairs Officer at 202-833-8773 ext. 211
    http://esa.sdsc.edu/jobs.htm *and*
    listjobs@esa.org

Educational Cyber Playground
    http://www.edu-cyberpg.com/Internet /SEARCHENGINES/JobList.html

Environmental Career Opportunities, P.O. Box 678, Stanardsville, VA 22973
    804-985-6529
    http://www.ecojobs.com *and*
    ecosubscriptions@mindspring.com

Environmental Job Opportunities, Institute for Environmental Studies, University of Wisconsin–Madison, 550 North Park St., 15 Science Hall, Madison, WI 53706
    608-263-1796
    http://www.ies.wisc.edu

Job Seeker, 24313 Destiny Ave., Tomah, WI 54660 (environmental job opportunities)
    608-378-4250
    http://www.tomah.com/jobseeker *and*
    jobseeker@tomah.com

Marine careers, New Hampshire Sea Grant College Program, Kingham Farm/ University of New Hampshire, Durham, NH 03824
    http://www.marinecareers.net/

Nature Conservancy, 4245 N. Fairfax Drive, Suite 100, Arlington, VA 22203
    800-628-6860
    http://nature.org/careers/ *and*
    comment@tnc.org

Opportunities in Environmental Careers. 1982. VGM Career Horizons: Lincolnwood, IL.
    4255 W. Touhy Ave., Lincolnwood, IL 60647

National Park Service
    http://www.nps.gov/personnel *and*
    http://www.usajobs.opm.gov/

National Wildlife Federation, 11100 Wildlife Center Drive, Reston, VA 20190
    800-822-9919
    http://www.nwf.org/jobopps/index.cfm *and* jobopp@nwf.org

Resumes for Environmental Careers. 1994. VGM Career Horizons: Lincolnwood, IL.
    4255 W. Touhy Ave., Lincolnwood, IL 60647

Science Jobs Page, American Association for the Advancement of Science, 1200 New York Ave. NW, Washington, DC 20005
    202-326-6417
    http://recruit.sciencemag.org/ *and*
    membership2@aaas.org

Sustainable Business Insider
    http://www.sustainablebusiness.com/ *and* rona@sustainablebusiness.com

Texas A&M University Jobs Page, Department of Wildlife and Fisheries Sciences, 210 Nagle Hall, Texas A&M University, 2258 TAMU, College Station, TX 77843
    979-845-5777
    http://wfscnet.tamu.edu/jobs.htm *and*
    http://wfscnet.tamucc.edu/email.html

If you skipped the Preface, note that the above list, as well as those lists that follow, includes sites that were familiar to the authors or sites brought to their attention. Please contact us at the address in the Preface if you have corrections, suggestions, or additions.

Chapter 2

A Crowded Planet

A central problem in disruption of ecosystems is having more organisms in an area than can be supported (the **carrying capacity**). When there are too many grasshoppers, lemmings, or rabbits in an area, the habitat becomes degraded. In the case of humans, the entire planet is involved. Wastes are turning up in "pristine" habitats such as open-ocean and arctic and antarctic areas; it is getting more difficult and expensive to find places to deposit our garbage; resources such as food and oil are in short supply in many areas; species of plants and animals are disappearing at a rapid rate; and global changes such as warming trends are a cause for concern.

Exponential Population Growth

Populations grow exponentially, and exponential growth can be very rapid. Lemmings, which are small rodents, have the potential to form a living rug of rodents on the landscape. This is true of all organisms including humans. The difference is that in humans the rate of increase is slower, with humans having to mature for more than a decade before being able to reproduce and having only one offspring every nine months. Phenomenal growth can be demonstrated by using a checkerboard and some pennies. If a penny is placed on square 1 and the number of pennies doubled from the square that precedes it, a table can be constructed showing the number of pennies for all 64 checkerboard squares. Table 2.1 gives some of the "milestones" in going from square 1 to square 64. Keep in mind that the human population now exceeds 6 billion and will soon reach 7 billion.

When populations become too dense, there are natural controls that retard further growth. Members experience more stress, are more susceptible to diseases, and become more aggressive, while

Table 2.1   *Using Checkerboard and Pennies to Illustrate Exponential Population Growth*

Square	Number of Stacked Pennies	Distance
1	1	Thickness of one penny
2	2	Thickness of two pennies
9	256	1 foot
21	1.04×10^6	1 mile
29	2.68×10^8	N.Y. to L.A.
48	1.41×10^{14}	Earth to sun
64	9.21×10^{18}	8.08 trillion miles of pennies, which is 1.3 light years from beginning to end

reproductive activity decreases as food, shelter, and space decline and conflict increases. The population may level off (**logistic growth**) or undergo a sharp decline and "crash."

Humans, at least in some parts of the world, are able to avoid a population crash by increasing the food supply, using modern medicines, and stabilizing population growth. Some populations may even reach zero population growth. However, long-term solutions to massive mortality involve limiting population growth, increasing or at least conserving our natural resources, and losing our **homocentric** (everything revolves around humans) viewpoint.

Human Manipulation of Population Size

Nonhumans

In smaller-sized species considered to be pests or "varmints," such as roof rats, humans try to eradicate as many individuals as possible. However, control is more often attained, and not complete extermination. With larger animals, population size is controlled by humans through **culling** (removal of a certain number of animals). More recently, birth control pills have been used for lions and some other species. In plants, culling is referred to as "thinning."

In both plant and animal populations, removal of individuals results in less competition and therefore more resources for other individuals. It is difficult for many people to realize that the removal of deer by shooting keeps the habitat from degrading and keeps it habitable for all species living there. Culling also keeps the remaining deer healthy and prevents them from reaching such high numbers that they may starve or die from other **density-dependent factors**—that is, the variables affected by the number of individuals found in a prescribed area. Sacrificing some members of the population also provides money and a free workforce in the form of recreational hunters for agencies that sell hunting

permits. Culling becomes even more controversial and emotional when the populations involved are elephants or other revered species.

Animals kept as companions by humans can also cause problems. Cats may be dropped off in an area where the former pet must then fend for itself. Also, some neighborhoods have high populations of cats that could never be supported in a natural environment. Often prey items include birds, lizards, rodents, and other species that contribute to biodiversity (see Chapter 4). Some suggestions for action include:

✓ Have cats, dogs, and other pets spayed

✓ Take strays to community animal control centers

✓ Volunteer and offer your services at humane societies

✓ Join a local environmental group

Humans

Population control has been instituted by the government in a few countries although this is not an option in free societies. In some countries, financial and material rewards are offered to those adhering to government-determined family sizes. These solutions highlight the seriousness of population problems in many countries. Without infringing on the rights of others, there are some things that help in controlling human population growth:

✓ Make people aware of overpopulation through presentations, articles, and conversation

✓ Organize humanitarian work in communities not as fortunate as yours

✓ Help family planning centers, which educate and provide advice on birth control methods

✓ Limit your own family size

✓ Be sexually responsible

Maintaining and Increasing Resources

Some species are able to recover their numbers relatively quickly only to become overcrowded and crash once again. This pattern of increase and decline is a population fluctuation. In the case of humans (*Homo sapiens*), who have few young and a long period of development with much parental care, recovery from a population crash can take a relatively long time. Long recovery of populations is common in **K-selected species**, those that have few young, a long period of development, and a lot of parental care. Other organisms have populations that can rebound quickly. Flies have many young, a short period of development, and little or no parental care, and are said to be **r-selected species**.

Common resources are often degraded because everyone tries to maximize their own benefit without any regard to future use of the resource (the **tragedy of the commons**). Many fisheries have been decimated by overfishing. Fishermen from many nations extracted as many fish as possible from fisheries located and shared in international waters, with no regard for the future condition of the fisheries and with profit the major consideration. As a result, some fish populations off of Peru and other places have been unable to recover and have become a lost resource for all nations.

Game management and conservation strategies try to avoid this needless loss of living resources. Excess individuals may be removed from the population by conservators and hand-reared. A good example is that of whooping cranes, which always produce two eggs but only one of which survives in the wild. Hand-rearing allows one egg to be removed, incubated, and cared for by humans. Hand-rearing of endangered condors has made it feasible to re-release condors into their former range.

Unlike minerals, oil, and other resources that are not renewable, living things have the capacity to reproduce using available resources. Populations thus have a certain number of individuals based on the carrying capacity of the habitat. Since a number of individuals beyond the carrying

capacity are not going to survive anyway, the excess individuals can be harvested without affecting the normal population size. This concept of **sustainability** (continuing yield) is like having a deposit in the bank and living off the interest.

Increasing yields from livestock is realized by adding food and shelter for animals. Likewise water and fertilizers can increase crop production, but there are problems in augmenting habitats with more resources, which are discussed in the next chapter. Genetic manipulation of plants and animals may increase yields, but there may be a problem with these modified genes potentially becoming incorporated in natural populations. Some of the consequences of genetic manipulation are straightforward such as seedless varieties of agricultural or horticultural products producing sterile plants, but many impacts are often unpredictable. Refer to the list of 50 harmful effects of genetically modified foods listed by CQS in the reference section at the end of this chapter.

Activities that will give practical experience and that should contribute to a sense of "doing the right thing" include:

✓ Volunteer or intern at fisheries and national parks

✓ Visit or volunteer at a local zoo, botanical garden, or park

✓ Grow some of your own food to supplement purchased groceries

✓ Become knowledgeable about certain crops that have been genetically manipulated

✓ Help in the recovery of nonhuman species affected by the expansion of human populations

✓ Assist in the recovery of impacted environments to their former condition

✓ Join local habitat restoration efforts, public education programs, and activities

Developing an Ecocentric Viewpoint

Perhaps the most important thing that can be done to solve population problems is to follow a **Land Ethic**: lose the **homocentric** (everything revolves around humans) viewpoint and become more **ecocentric** (realizing that humans are but a part of the vast complex of interacting living and non-living components of an ecosystem) as taught by Aldo Leopold.

A demonstration of this interconnectiveness involves **Closed Environmental Systems** (CESs). These systems are sealed, receiving no inputs and having no outputs; energy in the form of sunlight is the only exception. Plants take the solar energy in the form of light and photosynthesize food and produce oxygen. Animals utilize the foods and produce CO_2, which the plants utilize. Water and nutrients in the sediment circulate within the sphere and between the plants and animals. CESs are available for purchase as sealed glass spheres, but plastic soda containers are inexpensive and an easy way to construct a CES. Just add some water, plants, and simple animals such as shrimp, and screw the cap on to complete a miniature ecosystem.

As the space program offers increasingly detailed views of our planet in the lifeless void of space, we should become more aware of earth as a spaceship that needs to be maintained if we are to survive. The Land Ethic notes that we must not abuse the environment because we feel it belongs to us; instead we should view our surroundings as a community to which we each belong.

In order to succeed in having a non-homocentric viewpoint, it is important to have a sense of environmental ethics: a sense of construction must prevail over a sense of destruction; individuals must become less egocentric and more community minded; individual responsibility must be accepted for its part in some of the problems the planet is experiencing; and there must be recognition that the planet is not just for our generation but for all future generations.

References and Resources on Population Limits

Arcytech, Population Growth and Balance (hands-on activities) http://www.arcytech.org/java/population *and* jacobo@arcytech.org

Boulding, K.E. 1966. Economics of the coming spaceship earth. Pp. 3–14 in *Environmental Quality in a Growing Economy*. Henry Jarrett, ed. John Hopkins University Press: Baltimore, MD.

Cohen, Joel. 1996. Ten myths of populations. *Discover* 17: 42–47.

CQS, 2192 Massachusetts Ave., Cambridge,
MA 02140 (genetically modified foods)
617-491-7646
http://www.cqs.com/50harm.htm *and*
jon@cqs.com

DAPTF (Declining Amphibian Populations
Task Force)
http://www.open.ac.uk/daptf *and*
DATPF@open.ac.uk

Diamond, J. 1992. Must we shoot deer to
save nature? *Natural History*, August: 2–8.

Diamond, J. 1995. Easter's end. *Discover* 16:
62–69.

Ehrlich, Paul R. and Anne H. Ehrlich. 1990.
The Population Explosion. Doubleday:
New York, NY (For discussion, see
http://2think.org)

Family Planning Perspectives, The Alan
Guttmacher Institute, 102 Wall St.,
New York, NY 10005
212-248-1111
http://www.agi-usa.org/about/ *and*
info@agi-usa.org

Hardin, Garrett. 1975. Carrying capacity as
an ethical concept.
http://www.esva.net/~leo/carrycap.html

Leopold, Aldo. 1949. *A Sand County
Almanac*. Oxford University Press: New
York, NY (the Land Ethic)

NPG (Negative Population Growth),
1717 Massachusetts Ave. NW, Suite 101,
Washington, DC 20036
202-667-8950
http://www.npg.org *and* info@npg.org

Planned Parenthood Federation of America,
810 Seventh Ave., New York, NY 10019
800-230-PLAN *and* 212-541-7800 *and*
212-245-1845
http://www.plannedparenthood.org *and*
member.services@ppfa.org

Population Reference Bureau,
1875 Connecticut Ave. NW, Suite 520,
Washington, DC 20009
202-483-1100
http://www.popnet.org *and*
popref@prb.org

Shaffer, Jonathon A. 1993. Closed ecological systems. *Carolina Tips* 56: 13–15.

U.S. Census Bureau, 4700 Silver Hill Rd.,
Suitland, MD 20746 (Population Projections)
301-457-4608
http://www.census.gov/population/
www/projections/popproj.html *and*
comments@census.gov

(ZPG) Zero Population Growth, 1400 16th
St. NW, Suite 320, Washington, DC 20036
202-332-2200
http://zpg.org *and* info@zpg.org

Chapter 3

Wrenches in the Ecosystem Machinery

Large populations can alter the normal functioning of an ecosystem by disrupting the flow of nutrients and energy in food webs and by removing key species and altering abiotic (nonliving) components of the ecosystem.

What Is Pollution?

Pollution is anything that causes an undesirable change in an ecosystem that can negatively affect the health or activities of organisms or the characteristics of nonliving components (soil, air, water). Pollutants may be unsightly, deadly as in the case of toxic chemicals, or simply irritating as with some types of noise pollution. Disruptions may be undesirable in one aspect but beneficial in others. Fires, for example, release smoke and other substances that pollute the air, but they also return nutrients to the soil.

It has been said that pollution is "a resource out of place." Aluminum cans along a roadside are a form of pollution, although in a recycling bin the cans become a source of aluminum that will save on the energy needed to mine and refine aluminum ore. Ozone (O_3) is a pollutant in the workplace causing irritation to the eyes and respiratory tract, but in the upper atmosphere ozone plays a vital role in filtering out harmful ultraviolet radiation.

Not all pollution results from human activity. Large rafts of seaweed may wash up onto a beach, forming an unsightly, odiferous mass that pollutes the beach. From an ecologist's viewpoint, however, the seaweed has the desirable effect of providing nutrients that will feed the bacteria that serve as a food source for filter feeders such as clams, which in turn will serve as food for crabs, fishes, birds, and other species that are valued parts of a beach ecosystem.

Pollution can also be defined as the transfer of harmful materials to the environment by human activity. This definition emphasizes the amounts of materials, whether from **anthropogenic** (human-generated) or natural sources.

Because everything within an ecosystem is interrelated, the adjustments we make in one area of the ecosystem may result in changes in other areas of the ecosystem. Solving one problem may result in a problem somewhere else. Introducing mongooses on an island to get rid of rats may solve a rat problem, but there is a good chance that rats will not be the only species affected. Some island bird species, such as in Hawaii and Guam, have been decimated by introductions. As another example, wind energy is a desirable energy source because pollutants are not released, but rotating propellers may kill many migrating birds passing through the area. Solutions must be thoroughly researched, and even then there may be some unaccounted-for surprises and compromises to be considered.

Energy Flow, Habitats, and Impacts

The circulation of material and energy through and between ecosystems by organisms, geologic forces, and chemical processes is known as a **biogeochemical cycle**. These cycles may become restricted and may also circulate toxins. Such disruptions or impacts can have profound effects. Chlorinated hydrocarbons such as DDT do not break down easily and have an affinity for fatty tissues so that they are not easily excreted. As these persistent chemicals accumulate in an organism (especially filter feeders such as oysters), they also accumulate in greater amounts in the animals that eat oysters and in even higher levels in the animals that eat the oyster-eaters. This accumulation process is known as **biomagnification**.

The physiological response of organisms to these toxins varies. Brown pelicans with high DDT concentrations laid eggs with very thin shells that broke when a parent sat on them, whereas individuals from other species simply died. Other disruptive materials that become concentrated in living things include heavy metals such as lead, which had been commonplace in paints in years past, PCBs (polychlorinated biphenyls) found in some printing inks and plastics, and accumulation of radiation fallout such as occurred after the Chernobyl nuclear disaster.

Other problems arise from an overabundance of nutrients such as nitrogen and phosphorus entering waterways. Phosphorus was formerly prevalent in many detergents. Fertilizers having both nitrogen and phosphorus may also be washed into streams and lakes. As a result, algae that are normally limited in their growth by the absence of these nutrients may suddenly proliferate in great abundance and choke waterways. This explosion of growth is known as an algal bloom. As the nutrients are used up, the algae begin to die, and with decomposition the oxygen in the water is depleted. Eventually fish and other organisms die due to a lack of oxygen. The process is known as **cultural eutrophication**.

Emissions into the atmosphere are another problem. CFCs (chlorofluorocarbons) are gases in refrigerators and air conditioners that when released can destroy the ozone layer in the upper atmosphere. The ozone layer is important because it helps protect organisms from intense ultraviolet radiation. Sulfur, released when hydrocarbons such as petroleum are burned, can result in acid deposition and combine with precipitation to rain sulfuric acid. Aside from causing the slow degradation of limestone buildings and statuary, acid rain can have a devastating effect on vegetation and, by changing the pH of water, can inhibit fishes and other aquatic organisms from breeding successfully. Nitrogen oxides (NO_x) from automobile emissions are a group of chemicals that contribute to smog.

The **greenhouse gases**, which include CFCs, CO_2, ozone, methane, nitrous oxides, and water vapor, create a **greenhouse effect**. Instead of heat being radiated into space, heat becomes trapped close to the earth by the greenhouse gases, which then causes **global warming**. Increased atmospheric temperatures can change the species composition of ecosystems, affect food production, and raise the level of the sea. There is little argument that the earth is warming; there is considerable debate as to whether this warming trend is due to human activities or the same normal fluctuations in climate that brought us out of the Ice Age.

In addition to materials circulating in ecosystems, energy also flows through ecosystems. Solar energy is captured by plants, which are in turn eaten by herbivores, which are then eaten by carnivores.

The **second law of thermodynamics** notes that this transfer of energy is inefficient. At each step in this **pyramid of energy**, energy is lost by a factor of ten. If there are 100 kcal in the plants, there will be approximately 10 kcal available to the herbivore and only 1 kcal for the carnivore. One suggestion has been for humans to eat low on the food chain, thereby cutting out the middle cow and conserving energy that is lost through inefficiency. However, some essential amino acids and vitamins (B_{12}) are found in meat. Solutions are not always as simple as they appear.

Many environmental problems are the result of energy needs of humans. Not only is the rate of human population growth increasing, but the amount of energy used by each individual is also increasing. This highlights the need for more efficient use of energy resources, as opposed to simply exploiting more resources.

Solutions

As noted in Chapter 5 on involvement groups, concerned citizens should register to vote and influence legislators to base ecological decisions on sound scientific reasoning rather than economics, popularity, or political maneuvering. Compromises may have to be forged among spiritual, cultural, aesthetic, and other nonscientific considerations.

What else can be done? Citizens can become educated evaluators of scientific information and involved in monitoring waterways and cleanups. Write to companies with environmentally unsound products and explain to them why you will not buy their products. As changes in individual choices are made, so changes will have to occur in the manufacturing and production of goods. If a product is not being used, it will not be produced. Think ecologically in promoting a "new lifestyle for the new millennium."

It is important that people not feel helpless, especially now that the Internet provides rapid and inexpensive communication that can result in a groundswell of support. Purchasing power can influence environmentally oriented manufacturing strategies. Successful companies are those that listen to customers, so writing to inform them of your concerns is actually doing them a service.

It is easy to overlook some of the major success stories that many of us take for granted today. The ban of DDT in the 1960s was accomplished largely through the influence of Rachel Carson and her book *Silent Spring*. Some major fast food companies that formerly served their products in styrofoam containers no longer do so. The use of phosphorus in detergents has been dramatically reduced. Refrigerator and air conditioner manufacturers have changed coolants to reduce the emission of CFCs, and the use of non-leaded gasoline is now required along with the proper disposal of car batteries and tires. We have come a long way, and yet there are many things that can still be accomplished.

By decreasing, conserving, substituting, and modifying our needs, it is possible to reduce **ecological footprint** (an estimate of area required to support a population indefinitely). For more information and calculations of ecological footprints, refer to:

http://www.mec.ca/Apps/ecoCalc/ecoCalc.jsp

http://www.rprogress.org/programs/sustainability/ef/

http://www.lead.org/leadnet/footprint/default.htm

The checklists that follow will allow you to evaluate how well you are doing in keeping your ecological footprint small. The *Student Environmental Action Guide* highlights activities that have been particularly successful on campuses. There is also a Campus Ecology Program sponsored by the National Wildlife Federation. Particularly effective actions at work or school include:

✓ Conduct energy audits (and use fluorescent bulbs)

✓ Recycle paper, cardboard, glass, and aluminum

✓ Follow procedures for handling toxic materials

✓ Influence environmental course offerings

In looking at solutions, it is important to realize that sometimes there may not be better choices, only different ones. Dolphin-safe tuna are usually caught on long lines so that dolphins do not get caught in the nets and drown. However, seabirds may get caught on the long lines as well, so a choice must be made as to whether killing dolphins or seabirds is the lesser of two evils. All sides of an issue must be considered, not just the aspects that we find most appealing.

Checklist for Reducing Consumption of Resources

How many can you check off?

Water

Yard

☐ Use a broom rather than water to clean driveways and sidewalks

☐ Save rinse water (gray water) for watering plants and lawn

☐ Water lawns at dawn or dusk to reduce evaporation, and only when needed

☐ Keep lawns at least 2–3 inches high to keep from drying out or landscape without grass

☐ Xeriscape (plant species that require little watering)

☐ Use a mulching mower

☐ Use grass clippings and leaves for composting and providing mulch

☐ Mulch gardens to provide nutrients, prevent evaporation, and smother weeds

☐ Avoid paving large areas and terrace erodible slopes to reduce runoff of water

☐ Use a rain barrel or water tank to catch roof rainwater runoff for garden or car use

House

☐ Reduce water use (install faucet aerators and flow-restricting shower heads)

☐ Use low-flow toilets, water dams, or bottles filled with water to reduce tank capacity

☐ Check toilet leaks: add food coloring to tank; check bowl in 15 minutes for coloring

☐ Repair obvious leaks including leaking faucets

☐ Read water meter before and a half hour after turning off water-using appliances to detect leaks

☐ Do not let the water run when not using it (brushing teeth, dish washing)

☐ Use full washer loads

☐ Do not use a kitchen garbage disposal; compost food scraps if possible

Energy

General

☐ Look for energy-efficient apartments and houses

☐ Get an energy audit from the utility company for little cost or for free

☐ Organize an eco-olympics energy-saving competition between dorms

☐ Install light dimmer switches and motion sensors

☐ Turn off lights and appliances when not in use

☐ Keep refrigerators and freezers as full as possible to retain cooling

☐ Buy locally grown produce to save on transportation

☐ Eat low on the food chain

Temperature Control

☐ Opt for solar or on-demand water heaters and solar-powered calculators and watches

☐ Augment attic and roof insulation

☐ Insulate, caulk, weather-strip, and seal all leaks

☐ Insulate with storm doors and windows, window shades, and curtains

☐ Set the thermostat a little higher in summer and lower in winter

☐ Seal off rooms that are not in current use; temperature control used areas only

☐ Use timer on electric water heaters and set for 6–10 A.M. and 6–10 P.M.

☐ Turn the water heater to "low" (110–120°F); turn it off if vacationing

☐ Use a water heater insulating jacket

☐ Use cold water, not hot, for washing clothes, rinsing dishes, and other chores

☐ Plant vegetation around the house for cooling, and encourage others to plant

Appliances

☐ Refer to energy guide labels on appliances (refrigerators, air conditioners, heaters)

☐ Use a pressure cooker to save time and energy

☐ Use gas rather than electric appliances (stoves and heaters)

☐ Use manual versions of appliances (grinders, mixers, ice-cream makers)

☐ Use a clothesline (not a dryer); hand wash dishes and air dry them

☐ Do not use electric insect-zappers, which also kill many harmless insects

☐ Turn off appliances not in use

Other Consumables

Paper

☐ Use paper products made from recycled paper (the back is gray or tan if recycled)

☐ Use both sides of paper

☐ Use a rag instead of paper towels

☐ Use cloth instead of disposable diapers, but consider sanitation and bleach use

☐ Get off junk-mail mailing lists (contact the Direct Marketing Association)

Other

☐ Avoid plastics, especially nondurable plastics, which are petroleum-based

☐ Buy less meat = eat lower on the food chain

☐ In a buffet line, take only what you can consume

☐ Recycle glass, newspaper, cardboard, aluminum and other metals, plastics

☐ If a recycling program is not available in your area, start one

☐ Donate rather than throw away clothing and furniture (Goodwill, Salvation Army)

☐ Do not dispose of hazardous materials (solvents, oil, garden chemicals) down drains

☐ Note materials to be disposed of at a landfill or on hazardous waste collection day

Products from Living Things

☐ Honor the ban on ivory products to stop poaching

☐ Do not release helium balloons (sea turtles may die from eating fallen balloons)

☐ Avoid teakwood and other tropical hard-woods not from sustainable tree farms

☐ Do not purchase endangered species products (coral) or some exotic pets

Checklist for Reducing Output of Wastes and Pollutants

Because a car is the biggest polluter (con-tributing its weight in carbon oxides each year) and resource eater that most people will own, this is a good place to begin.

Car

Car Use

☐ Use the telephone rather than the car to shop for an item

☐ Walk, cycle, carpool, and use mass transit

☐ Choose a small, fuel-efficient car and skip automatic/power options

☐ Consider buying an electric/gasoline hybrid car

☐ Keep within the speed limit (40 mph gives best mileage); avoid quick stops and starts

☐ After one minute of engine idling, turn the engine off to stop emissions

Car Care

☐ Reduce noise pollution: avoid use of horn and keep the muffler and tailpipe in working order

☐ Have antipollution devices checked regularly

☐ Keep the engine well tuned (once a year or every 15,000 miles) to reduce emissions

☐ Change oil and air filters at regular intervals for efficiency

Car Products

☐ Use unleaded gasoline (now required in the U.S.)

☐ Never pour brake and transmission flu-ids, oil, and antifreeze onto the ground

☐ Recycle dead batteries and old tires

☐ Carry a trash bag

☐ Support local ordinances against overuse of signs along roadways

Purchasing Practices

☐ Do not buy products with excess packaging

☐ Buy products in reusable containers, or at least recyclable containers

☐ Avoid multimaterial packaging such as fruit juice boxes (they are not recyclable)

☐ Bulk purchase such nonperishables as rice, beans, and flour

☐ Buy durable goods rather than dispos-ables, and eat at places that use durable utensils

☐ Use a permanent razor with exchange-able blades rather than a plastic dispos-able

☐ Bring a container for take-outs or request that leftovers be packaged in wax paper

☐ Use plastic butter tubs and other con-tainers for storing leftovers rather than discard them

☐ Take your own large canvas, net, or paper bag for transporting purchases

Gifts

Purchase gifts manufactured from recyclable or renewable resources, and that help promote environmental awareness, such as:

- ☐ Recycled paper (stationery, cards, wrapping paper)
- ☐ Wooden toys (blocks, animals) rather than plastic
- ☐ Gift certificates for trees, shrubs, bird feeders, compost bins from a local nursery
- ☐ A basket of low-impact cleaners, detergents, and storage bags (Eco-Choice)
- ☐ Canvas shopping bags (Greenpeace, Sierra Club, the Nature Conservancy)
- ☐ T-shirts with nature-oriented designs
- ☐ Magazine subscription to *Ranger Rick* (ages 5–10), National Wildlife Federation
- ☐ Subscription to *My Big Backyard* (ages 2–4), National Wildlife Federation
- ☐ Subscription to *Sierra*, *Audubon*, *National Wildlife*, or similar magazines for adults
- ☐ An acre of tropical rain forest in Belize (Programme for Belize) in order to protect it

Replacing or Reducing Release of Materials That Pollute

Alternate forms of energy to oil include human power (doing things by hand rather than by machine), solar, wind, water, tidal, nuclear, and geothermal energy. Alternatives to toxic chemicals include hand weeding, crop rotation, mixed planting, release of sterile individuals and natural predators into the population of pests, and pheromone traps.

General

- ☐ Buy paper products unbleached or bleached without chlorine; dioxins are toxic
- ☐ Use low-toxicity paint strippers, and latex in place of oil-based paints
- ☐ Be aware that marine antifouling paints with organotins may damage marine ecosystems
- ☐ Put litter in litter baskets, and lead your dog off the sidewalk when it needs to evacuate
- ☐ Scale down the quantities needed in chemistry laboratory experiments (microscale)
- ☐ Buy domestic or locally grown foods in season (avoiding transportation costs)
- ☐ Be aware that irradiated foods may support the production of radioactive waste
- ☐ Familiarize yourself with ordinances regarding pollution, and report offenders
- ☐ Make posters that list and illustrate products that pollute or destroy environmental resources

Air

- ☐ Buy products in pump bottles rather than aerosols
- ☐ Avoid CFCs (chlorofluorocarbons) in aerosols (horns, cleaners, cleaning sprays)
- ☐ Recognize that styrofoam (polystyrene foam in cups, packing, trays) uses plastic and may emit CFCs
- ☐ Do not buy halon-releasing products (halon fire extinguishers) that act like CFCs
- ☐ Use regular, not self-lighting, charcoal due to the release of more hydrocarbons
- ☐ Protect against grass or forest fires, which can cause air and water pollution
- ☐ Give up smoking and convince a friend to do likewise
- ☐ Be aware that carbon tetrachloride in dry cleaning eats the ozone and that volatile hydrocarbons form smog

Plastic

- ☐ Avoid nondurable plastics; pollutants are emitted during manufacturing and burning
- ☐ Note that plastic yokes from six-pack cans kill and maim turtles, birds, and other wildlife
- ☐ Be aware that "biodegradable" plastics only fragment, produce pollutants, and are not recyclable
- ☐ Use a garbage can instead of plastic garbage bags

The Household

- ☐ Buy a plunger or plumber's snake rather than chemical drain cleaners
- ☐ Use the least toxic cleaners (phosphate-free detergents and nonchlorinated cleaners)
- ☐ Buy rechargeable batteries and a battery recharger
- ☐ Avoid synthetics (carpets, other flooring, and furniture upholstery)
- ☐ Check home heating annually for operating efficiency
- ☐ Replace pesticides with soaps (aphids), diatomaceous earth (slugs), Bt (caterpillars)
- ☐ Use compost in place of artificial fertilizers; do not burn leaves

Addresses for Environmentally Friendly Products

Supporting environmentally friendly products helps to solve many of the problems plaguing our planet. Below are some sources for buying responsibly.

Bio-Pax, Diversified Packaging Products, Inc. (boxes, cushioning, tape wrappers), MS 4001, 1265 Pine Hill Drive, Annapolis, MD 21401
    410-974-4411
    recycling+@andrew

Conservatree Paper Co., 100 Second Ave., San Francisco, CA 94118
    415-721-4230
    http://www.conservatree.com *and* paper@conservatree.com

Container Recycling Institute,
North 1911 Ft. Meyer Drive, Suite 702,
Arlington, VA 22209
    703-276-9800
    http://www.container-recycling.org/ *and*
    CRI@Container-Recycling.org

Co-op America (an alternative marketplace), 1612 K St. NW, Suite 600,
Washington, DC 20006
    800-58-GREEN
    http://www.coopamerica.com

Earth Care Products of Long Island,
200 E. Second St., Suite 8, Huntington
Station, NY 11746
    800-445-4445
    http://www.interstor.com/earthcare/ *and*
    earthcare@spec.net

Eco-Choice Business Directory, 513 N. Eglin
Parkway, Ft. Walton Beach, FL 32547
    800-449-8684
    http://users.pipeline.com.au/ink/wcm/
    gcindex1.htm *and* http://www.esalc.com

EcoMall (*Green Shopping Magazine; EcoMall
News*)
    http://www.ecomall.com/

Environmentally Preferable Products and
Services Vendor List. OA/DPMM–MO.
STATE RECYCLING PROGRAM
    http://www.oa.state.mo.us/purch/
    vendors.pdf

FREE (Fund for Renewable Energy and the
Environment), 101 Connecticut Ave. NW,
Suite 638, Washington, DC 20036
    202-466-6880

Gardens Alive (Natural Gardening
Research Center), 5100 Schenley Place,
Lawrenceburg, IN 47025
    812-537-8650

http://www.gardensalive.com *and*
gardenhelp@gardensalive.com

Green Culture, P.O. Box 1684, Laguna
Beach, CA 92652
    800-233-8438 or 949-643-8795
    http://www.greenculture.com/

Livos Phytochemistry of America, Inc.,
P.O. Box 1740, Mashpee, MA 02649 (paints,
stains, and wood preservatives without
toxic fumes)
    phone: 508-477-7955
    fax: 508-477-7988
    http://www.livos.com *and*
    info@livos.com

Mountainstar, P.O. Box 629, Lake Placid, NY
12946 (formerly Green Star General Store)
    518-891-8061
    http://www.gsgs.com *and*
    gsgs@stardesign.com

Natural Choice Directory, P.O. Box 18646,
Seattle, WA 98118
    206-722-4288
    http://naturalchoice.net *and*
    ncd@naturalchoice.net

Programme for Belize, #1 Eyre St., P.O. Box
749, Belize City, Belize (conserving tropical
rain forest)
    501-2-75616
    http://www.pfbelize.org/book.html *and*
    pfbel@btl.net

Recycler's World, RecycleNet Corporation,
P.O. Box 24017, Guelph, Ontario,
Canada N1E 6V8
    703-683-9025
    http://www.recycle.net

Seventh Generation (a wide variety of
products), 5801 Beacon St., Suite #2,
Pittsburgh, PA 15217

888-59-EARTH
http://www.greenmarketplace.com *and*
support@GreenMarketplace.com

Student Environmental Action Coalition
(SEAC), P.O. Box 31909, Philadelphia,
PA 19104
  215-222-4711
  http://www.seac.org/ *and* seac@seac.org

Addresses for Information on Environmentally Friendly Products

Requests can be made for catalogs or the
actual publication (fees may be required for
both). Be aware that many environmentally friendly products may cost more than
regular products.

American Council for an Energy-Efficient
Economy, 1001 Connecticut Ave. NW, Suite
801, Washington, DC 20036 (booklets: *The
Most Energy Efficient Appliances* and *Saving
Money with Home Appliances*)
  202-429-8873
  http://www.aceee.org *and*
  info@aceee.org

Council on Economic Priorities, 30 Irving
Place, New York, NY 10003 (guide: *Shopping
for a Better World*)
  800-729-4237
  http://www.cepnyc.org *and*
  info@cepnyc.org

Environmental Defense Fund, 257 Park
Ave. South, New York, NY 10010 (booklet:
*Protecting the Ozone Layer: What You Can
Do*)
  800-684-3322
  http://www.edf.org

Ethical Shopper (acquired by the Green
Marketplace), 5801 Beacon St., Suite 2,
Pittsburgh, PA 15217
  800-59-EARTH or 412-420-6400
  EthicalShopper.com *and*
  GreenMarketplace.com *and*
  support@EthicalShopper.com

Greenpeace (this action-oriented group is
one of the largest and best-known organizations) 702 H St. NW, Suite 300, Washington, DC 20001 (factsheet for alternatives to
toxic products: *Stepping Lightly on the Earth:
Everyone's Guide to Toxics in the Home*)
  800-326-0959
  http://www.greenpeace.org/ *and* inter@ct

Meadowbrook Press, 18318 Minnetonka
Blvd., Deephaven, MN 55391 (guide: *Shopping for a Better Environment*)
  http://www.meadowbrookpress.com

National Association of Diaper Services
(list of diaper services in your area),
994 Old Eagle School Road, #1019,
Philadelphia, PA 19087
  610-971-4850
  http://www.diapernet.com *and*
  jashiffert@multiservicemgmt.com

Pennsylvania Resources Council, Environmental Living Center, 3606 Providence
Road, Newton Square, PA 19073 (handbook: *Become an Environmental Shopper*)
  610-353-1555
  http://www.prc.org

RAN (Rainforest Action Network),
221 Pine St., Suite 500, San Francisco, CA
94104 (factsheet: *Tropical Timber Factsheet*)
  415-398-4404
  http://www.ran.org/ran/ *and*
  rainforest@ran.org

Tilden Press, Inc., 6 Hillwood Pl., Oakland, CA 94610 (*The Green Business Newsletter* and *The Green Consumer* (1990) book by J. Elkington, J. Hailes, and J. Makower)
> 800-955-GREEN
> http://www.greenbizletter.com *and* gbl@greenbiz.com

World Wildlife Fund, Traffic (USA), 1250 24th St. NW, Washington, DC 20037 (guide: *Buyer Beware* provides information on illegal wildlife products)
> 800-634-4444
> http://www.worldwildlife.org

Addresses for Information on Saving Energy and Avoiding Pollution

American Solar Energy Society, 2400 Central Ave., Suite G-1, Boulder, CO 80301 (*Solar Today*)
> 303-443-3130
> http://www.ases.org/ *and* ases@ases.org

American Wind Energy Association, 122 C St. NW, Suite 380, Washington, DC 20001
> 202-383-2500
> http://www.awea.org *and* windmail@awea.org

CAREIRS (Conservation and Renewable Energy Inquiry and Referral Service), P.O. Box 8900, Silver Spring, MD 20907 (tips on energy conservation)
> 800-523-2929

Direct Marketing Association, 1120 Avenue of the Americas, New York, NY 10036
> 212-768-7277
> http://www.the-dma.org *and*

privacy@the-dma.org

Energy Efficiency and Renewable Energy Network (EREN), U.S. Department of Energy
> http://www.eren.doe.gov/ *and* kevin_eber@nrel.gov

Environmental Defense Fund (brochure: *Coming Full Circle, Successful Recycling Today*), 257 Park Ave. South, New York, NY 10010
> 212-505-2100
> http://www.edf.org

Gardener's Supply, 128 Intervale Road, Burlington, VT 05401 (brochure: *Simple Steps to Successful Composting*)
> 888-833-1412
> http://www.gardeners.com *and* info@gardeners.com

Global Releaf (American Forestry Association), P.O. Box 2000, Washington, DC 20013 (reforestation project)
> 202-955-4500
> http://www.americanforests.org/ global_releaf/ *and* info@amfor.org

Household Hazardous Waste, 1031 E. Battlefield, Suite 224-B, Springfield, MO 65807
> 417-889-5000
> http://outreach.missouri.edu/owm/hhw *and* owm@missouri.edu

NATAS (National Appropriate Technology Assistance Service), P.O. Box 2525, Butte, MT 59702 (tips on energy conservation)
> 800-428-1718
> http://www.homestead.org/enerorgs.htm

National Recycling Coalition, USA, 1727 King St., Suite 105, Alexandria, VA 22314
> 703-683-9025

http://www.recycle.net/recycle/ Associations/rs000145.html *and* darrick@homestead.org

National Wildlife Federation, Water Resources Program, 1400 Sixteenth St. NW, Washington, DC 20036 (guide: *Citizen's Guide to Water Conservation*; plans for building barn owl, bluebird and bat houses; Energy Savers' Source List)
>   800-822-9919
>   http://www.nwf.org/campusecology/ *and* http://www.pueblo.gsa.gov/cic-text/ housing/energy-savers/sourcelist.html

Rocky Mountain Institute, 1739 Snowmass Creek Road, Snowmass, CO 81654 (conservation research)
>   970-927-3851
>   http://www.rmi.org *and* outreach@rmi.org

Worldwise, Inc., Department WS, P.O. Box 3360, San Rafael, CA 94912 (sustainability advisor)
>   415-721-7400
>   http://store/yahoo/com/worldwide/ deblyndadsus.html *and* info@worldwise.com

General Guides to Being Environmentally Responsible

Causey, Ann S. 1991. *Environmental Action Guide. Action for a Sustainable Future.* Benjamin/Cummings Publishing: Redwood City, CA.

Lamb, Marjorie. 1990. *2 Minutes a Day for a Greener Planet.* Harper & Row Publishers: San Francisco, CA.

MacEachern, Diane. 1995. *Save Our Planet: 750 Everyday Ways You Can Help Clean Up the Earth.* Dell: New York, NY.

The Earthworks Group. 1989. *50 Simple Things You Can Do to Save the Planet.* Earth Works Press: Berkeley, CA.

Student Environmental Action Coalition. 1991. *The Student Environmental Action Guide: 25 Simple Things We Can Do.* HarperCollins/The Earth Works Group. (SEAC, P.O. Box 31909, Philadelphia, PA 19104)

Wackernagel, Mathis and William E. Rees. 1996. *Our Ecological Footprint: Reducing Human Impact on the Earth.* New Society Publishers: Gabriola Island, B.C., Canada.

Selected References for Combating Pollution and Waste

Heathcote, Willimas. 1991. *Autogeddon.* Jonathon Cape: London, England.

MacKenzie, James J. et al. 1992. *The Going Rate: What It Really Costs to Drive.* World Resources Institute: Washington, DC.

Makower, Joel. 1992. *The Green Commuter.* National Press: Washington, DC.

Chapter 4

Habitat Diversity for a Quality Life

If we conserve our existing habitats and do not permanently consume or damage our limited resources, then we will not have to restore or replace (if either is possible) natural areas. Quantitative monitoring of natural areas is important because it provides a baseline for measurement of how fast, how far, and in what manner changes are being made to natural ecosystems. Also, by being aware of what we have in the way of natural resources and the cost of cleaning up our earlier mistakes (refer to the literature on Superfund cleanups), the less likely it is that we will foul our own nest.

Biodiversity

Biodiverse systems have many different species, whereas **monocultures** are characterized by only one species. Hundreds of miles of one species of grain or a pine tree plantation can be boring and aesthetically less desirable. Also, such monocultures have a greater risk of experiencing widespread disease and pests that make these areas inherently more unstable. Biodiverse ecosystems tend toward stability because if a part of the ecosystem puzzle becomes lost there may be other replacement species that can fill the role of the missing part.

Diversity is in part a consequence of **zonation**. There may be a gradation of change, such as when the habitat becomes dryer with elevation or distance away from a water source. Different areas or zones along this gradation have different plants and animals adapted to particular conditions that range from wet to dry or other variables such as shade to light.

However, serious damage can be caused by trying to make an area more biodiverse by the introduction of species not native to the area (**invasives**). Without natural controls to population growth, rabbits have been able to overrun Australia, zebra mussels have become a pest in the waters of North America, and species such as the brown tree snake are a threat to cross borders into North America. There was also the inadvertent introduction of smallpox to North America by the colonists, which devastated native human populations that had not developed an adequate immune defense against the disease.

On a global scale, one can deliberate what the outcome would be of building a sea-level Panama Canal that would connect the Pacific Ocean and the Caribbean Sea. Many native species suffer from the introduction of **exotic** (foreign) species due to the **competitive exclusion principle**: plants or animals with the same requirements cannot coexist in the same area due to competition.

Identification of species is important in being able to describe what **biota** (living things) are found in an area; only then can it be determined what has been impacted or lost altogether. Listing a "rat" is not helpful since there are hundreds of species, some of which are serious pests and others which are endangered.

A measurement of diversity in an area can be calculated mathematically by the use of biodiversity indices. Shannon's Index is a well-known measure of biodiversity that involves the degree of probability that the next sample taken will be the same as or different from the previous sample taken. If the probability is that the sample will be different, then there is a higher indication of biodiversity.

Genetic diversity is currently being lost through selective breeding and through extinction. Alligators, whooping cranes, brown pelicans, and oryxes were all on the road to extinction, but have recently made progress toward recovery. To guard against loss of genes that have evolved over millions of years, seed banks and frozen tissue banks have become the safe deposit boxes of diversity.

Some activities to promote biodiversity include:

✓ Make a species list of plants and animals found in your community

✓ Make a list of species that have been introduced into your area

✓ Locate fossil sites near you and compare past and present faunas

✓ Plant to attract butterflies, birds, and other animals

✓ Provide barn owl, bluebird, and bat houses

✓ Put decals or stickers on your windows to keep birds from flying into them

✓ Avoid certain genetically manipulated seeds and produce (refer to Greenpeace list below)

✓ Work with groups such as garden clubs to landscape areas in need of greenery

✓ Support zoos and captive breeding programs

✓ Do not buy exotic animals that are threatened in the wild; buy from captive breeders

✓ Use plants propagated in nurseries rather than plants taken from wild areas

Keys and Identification Aids

No one person can be an expert on the identification of all living things. Nonetheless, it is important to understand what species are living in particular areas because they may be endangered or simply desirable species to conserve. Baseline data is required if changes occur even to the most common species.

There are relatively inexpensive field guides that help in the identification of species. *The Pictured Nature Key Series* published by Wm. C. Brown (Dubuque, IA) offers a series of identification manuals, including aquatic insects, aquatic plants, freshwater algae, fishes, protozoa, birds, and mammals. Golden Press (New York, NY) has published a series of identification guides that include well-illustrated keys in color.

Identification can also be accomplished through the recognition of field marks or specific characteristics without going through the process of using a key. The *Peterson Field Guide Series* published by Houghton Mifflin (Boston, MA) is useful for covering identification of a wide number of ecological components, including plants, minerals, shells, butterflies and other insects, vertebrates, and the atmosphere. The Audubon Society has a series of field guides with excellent photographs published by Alfred A. Knopf (New York, NY).

The World Guide to Bird Guides by G.C. Hickman and G.L. Maclean (1995) lists available field guides to birds by country and is available for $13 on disc from G. Hickman, 6606 Sahara Drive, Corpus Christi, TX 78412.

Some of the scientific names can be long and difficult to remember. However, if some of the common prefixes and suffixes are known, the terms can be learned rather than memorized. A good source for these word-building forms is:

Borror, Donald J. 1960. *Dictionary of Word Roots and Combining Forms. Compiled from the Greek, Latin, and Other Languages, with Special Reference to Biological Terms and Scientific Names*. Mayfield Publishing: Mountain View, CA.

Biomes and Habitat Protection

Rather than trying to save only threatened species, it is better to preserve entire ecosystems. Larger conservation areas are almost always preferred over small areas, so that **fragmentation** of areas can be avoided with the understanding that in some cases it may be necessary to have separate conservation areas to prevent all the "eggs" being put into one basket.

In most cases involving separate areas, however, two halves do not make a whole. A road bisecting a conservation area will reduce population size and gene diversity because many animals will not cross the pavement. One solution to this problem is to have a corridor of natural habitat linking the two distinct areas. The habitat area also reduces fragmentation because the edges of the area have more wind and light and higher temperatures than are found in the central portion of the area. Nonetheless, circular reserves may not be optimal where boundaries can extend out from the perimeter to include such desirable features as a lake or river. Decisions on nature reserves must also take into account political boundaries, economics, and social factors.

Some Diverse Areas on Our Planet and Their Particular Problems

Deserts

Arid regions are becoming more extensive as freshwater becomes increasingly sparse  due to a process known as desertification. Because these areas are not very productive, species may be few. The harshness of the environment also puts stress on species that are hunted such as the Arabian oryx.

Grasslands

The loss of grasslands has been caused largely through plowing and overgrazing. Some grasslands require periodic fires to remain grasslands and prevent encroachment by woody growth. Some of the North American Indians started fires to keep the grasslands in vigorous growth, which provided food for the buffalo, a source of meat and hides for the Indians.

Shrublands

Areas dominated by shrubs, such as chaparral, are not as common as some of the other biomes. Fire is also a common feature of many of these areas.

Forests

Forested areas have been cleared for agriculture and logged for wood. This defor-estation represents a loss of areas that acted as carbon dioxide (CO_2) sinks, which could remove CO_2 from the atmosphere and incorporate it in living tissue through photosynthesis. This reduction in the removal of CO_2 is unfortunate because CO_2 is a greenhouse gas that promotes global warming. Some of the oldest living species on earth are trees that also provide shade and shelter and anchor the soil.

The loss of tropical rain forests is particularly damaging, since rain forests do not recover to their former grandeur because nutrients are stored in the living tissue of the forest and the soils are nutrient poor. Even more disturbing is the fact that much of our planet's biodiversity resides in rain forests and that organisms are going extinct without ever having been seen by humans. At the present rate of destruction, all our rain forests will be gone in 25 years.

Tundra

Pipeline construction can cause fragmentation in the range of some tundra species such as caribou. There have also been problems with biomagnification, with radioactive fallout being concentrated in lichens and then eaten by caribou.

Wetlands

Although our planet is covered by water, only 3 percent is freshwater found in such areas as swamps, marshes, bogs, lakes, and rivers. Before wetlands were appreciated for their ability to remove pollutants and serve as nurseries for fishes and other organisms, many were lost by being filled in with soil for building space. The Adopt-a-Wetland

Program provides educational information and is involved in conservation work.

Rivers, Streams, Lakes, and Ponds

Modified because of industry, recreation, and protection of homes, rivers have suffered from the construction of levees. Levees stop flooding, but they also prevent the deposition of nutrient-rich soils on the floodplain. Dams change whole ecological communities due to changes in the depth and temperature of the water. Beaver dams are not as grandiose as human dams but still have a profound effect on some ecosystems.

Marine Habitats

Some areas of oceans and seas have been overfished. Corals and some of the more brightly colored inhabitants have decreased due to collection for home aquaria and to damage caused by boats anchoring in these areas for scuba diving. Oil spills continue to threaten coastlines, and some whales, dolphins, and porpoises continue to be stranded on beaches. The Marine Mammal Stranding Network assists in helping to save these animals.

Islands

The introduction of exotic species to islands has resulted in the devastation and extinc-tion of the natural plants and animals. Oceanic islands are important because they help answer "what if . . ." questions such as "What if species evolved without the presence of carnivores?" The study of islands also provides important data on how habitat islands such as game reserve areas should be designed and maintained.

Organizations for Preserving Biodiversity

Other organizations may be listed in the "Directories" section in Chapter 6.

American Zoo and Aquarium Association (AZA), 8403 Colesville Rd., Suite 710, Silver Spring, MD 20910
 301-562-0777
 http://www.aza.org/ *and* membership@aza.org

Bat Conservation International, P.O. Box 162603, Austin, TX 78716
 800-538-BATS
 http://www.batcon.org

Caribbean Conservation Corporation, 4424 NW 13th St., Suite A1, Gainesville, FL 32609
 800-678-7853
 http://www.cccturtle.org *and* ccc@cccturtle.org

Center for Marine Conservation, 1725 De Sales St. NW, Suite 600, Washington, DC 20036
 202-429-5609

Center for Plant Conservation, P.O. Box 299, St. Louis, MO 63166
 314-577-5100
 http://www.mobot.org

Conservation Foundation, Kensington Gore, London SW7 2AR, UK
 44 20 7591 3111
 http://www.conservationfoundation.co.uk *and* conservationf@gn.apc.org

Conservation Fund, 1800 N. Kent St.,
Suite 1120, Arlington, VA 22209
  703-525-6300
  http://www.conservationfund.org

Conservation International, 1919 M St.
NW, Suite 600, Washington, DC 20036
  202-912-1000
  http://www.conservation.org

Defenders of Wildlife, 1101 14th St. NW,
#1400, Washington, DC 20005
  202-682-9400
  http://www.defenders.org/

EPA (Environmental Protection Agency),
1200 Pennsylvania Ave. NW, Washington,
DC 20460
  http://www.epa.gov/maia/html/
  intro-species.html

Greenpeace, 702 H St. NW, Washington,
DC 20001
  800-326-0959
  http://www.greenpeace.org/~geneng/
  and inter@ct

National Audubon Society, 700 Broadway,
New York, NY 10003
  212-979-3000
  http://www.audubon.org *and*
  join@audubon.org

National Conservation Training Center,
USFWS, Rt. 1, Box 166, Shepherdstown,
WV 25443
  http://training.fws.gov

National Wildlife Federation, 11100
Wildlife Center Drive, Reston, VA 20190
  703-438-6000
  http://www.nwf.org/

Nature Conservancy, 4245 N. Fairfax Drive,
Suite 100, Arlington, VA 22203

800-628-6860
http://nature.org *and* comment@tnc.org

NBII (National Biological Information
Infrastructure), National Program Office,
USGS Biological Informatics Office,
302 National Center, Reston, VA 20192
  703-648-6244
  http://www.nbii.gov/index.html *and*
  http://www.invasivespecies.gov/profile/
  bts.shtml

Oceanic Society, Fort Mason Center,
San Francisco, CA 94123
  800-326-7491

Plant Conservation Alliance, Bureau of
Land Management, 1849 C St. NW,
LSB-204, Washington, DC 20240
  202-452-0392
  http://www.nps.gov/plants/ *and*
  olivia_kwong@blm.gov

Predator Conservation Alliance, P.O. Box
6733, Bozeman, MT 59771
  406-587-3389
  http://www.predatorconservation.org
  and pca@predatorconservation.org

Rainforest Action Network, 221 Pine Street,
Suite 500, San Francisco, CA 94104
  415-398-4404
  http://www.ran.org/ran/ *and*
  rainforest@ran.org

Reef Relief, P.O. Box 430, 201 William St.,
Key West, FL 33041
  305-294-3100
  http://www.reefrelief.org *and*
  reef@bellsouth.net

Save America's Forests, 4 Library Court SE,
Washington, DC 20003
  202-544-9219
  http://www.saveamericasforests.org/

Seed Savers Exchange, 3076 North Winn Road, Decorah, IA 52101
> http://www.seedsavers.org

Sierra Club, 730 Polk St., San Francisco, CA 94109 or 85 Second St., San Francisco, CA 94105
> 415-977-5500
> http://www.sierraclub.org *and* information@sierraclub.org

Society for Conservation Biology, Box 351800, University of Washington, Seattle, WA 98195
> 206-616-4054

Society for Conservation GIS, P.O. Box 661458, Los Angeles, CA 90066
> http://www.scgis.org *and* lanius@rocketmail.com

TreePeople, 12601 Mulholland Drive, Beverly Hills, CA 90210
> 818-753-4600
> http://www.treepeople.org *and* mskerrett@treepeople.org

U.S. Fish and Wildlife Service (refer to web page for e-mail directories)
> http://www.fws.gov/ *and* http://endangered.fws.gov/wildlife/ html#species *and* contact@fws.gov

Whale and Dolphin Conservation Society (WDCS), Alexnder House, James St., West Bath BA1 28T, UK
> 44-0-1225-35433 (outside of UK)
> http://www.wdcs.org *and* info@wdcs.org

Wildlife Conservation Society, 2300 Southern Boulevard, Bronx, NY 10460
> 718-220-5100

http://wcs.org *and* membership@wcs.org

World Wildlife Fund (protection of wildlife and habitats, focusing on third world countries), 1250 24th St. NW, P.O. Box 97180, Washington, DC 20090
> 800-CALL-WWF
> http://www.worldwildlife.org

Zero Population Growth (voluntary family planning to balance resources/population), 1400 16th St. NW, Suite 320, Washington, DC 20036
> 800-POP-1956 or 202-332-2200
> http://www.zpg.org *and* info@zpg.org

References for Preserving Biodiversity

General

Berger, J.T. 1990. *Environmental Restoration*. Island Press: Washington, DC.

Carson, Rachel. 1962. *Silent Spring*. Houghton Mifflin: Boston, MA.

Clapham, W.B., Jr. 1984. *Natural Ecosystems*. Macmillan: New York, NY.

Cox, George W. 1999. *Alien Species in North America and Hawaii*. Island Press: Covelo, CA.

Fielder, P.L. and S. Jain, eds. 1992. *Conservation Biology*. Chapman & Hall: New York, NY.

Fincham, J.R.S. and J.R. Ravitz. 1991. *Genetically Engineered Organisms: Benefits and Risks*. John Wiley: New York, NY.

Hardin, G. 1968. The tragedy of the commons. *Science* 162: 1243–1248.

Meffe, G.K. and C.R. Carroll. 1994. *Principles of Conservation Biology*. Sinauer: Sunderland, MA.

Nilson, Richard, ed. 1991. *Helping Nature Heal: An Introduction to Environmental Restoration*. Ten Speed Press: Berkeley, CA.

Premack, R.B. 1993. *Essentials of Conservation Biology*. Sinauer: Sunderland, MA.

Reaka-Kudla, Marjorie L., Don E. Wilson, and Edward O. Wilson, eds. *Biodiversity II: Understanding and Protecting Our Natural Resources*. Joseph Henry Press: Washington, DC.

Vitousek, P.M., C.M. D'Antonio, L.L. Loope, and R. Westbrooks. 1999. Biological invasions as global change. Pp. 218–228 in Peter Kareiva, ed., *Exploring Ecology and Its Applications*. Readings from Scientific American. Sinauer Associates: Sunderland, MA.

Wilson, E.O., ed. 1988. *Biodiversity*. National Academy Press: Washington, DC.

Wilson, E.O. 1992. *The Diversity of Life*. Harvard University Press: Cambridge, MA.

Areas

Brown, Lauren. 1985. *Grasslands*. Random House: New York, NY.

Goldman-Carter, Jan. 1989. *A Citizen's Guide to Protecting Wetlands*. National Wildlife Federation: Washington, DC.

Lauw, G.N. and M.K. Seely. 1982. *Ecology of Desert Organisms*. Longman: New York, NY.

Mabberly, D.J. 1983. *Tropical Rainforest Ecology*. Blackie: London, England.

MacMahon, James A. 1985. *Deserts*. Random House: New York, NY.

Mitsch, W.J. and J.G. Gosselink. 2000. *Wetlands*. Van Nostrand Reinhold: New York, NY.

Noss, Reed F., ed. 1999. *The Redwood Forest: History, Ecology, and Conservation of the Coast Redwoods*. Island Press: Covelo, CA.

Scott, J. Michael, Sheila Conant, and Charles van Riper, III, eds. 2000. *Evolution, Ecology, Conservation, and Management of Hawaiian Birds: A Vanishing Avifauna*. Cooper Ornithological Society: Camarillo, CA.

Thorne-Miller, B. and J. Catena. 1990. *The Living Ocean: Understanding and Protecting Marine Diversity*. Island Press: Covelo, CA.

Yates, Steve. 1989. *Adopting a Wetland*. Adopt-a-Stream Foundation: Everett, WA.

Species Accounts

Doughty, Robin W. 1989. *Return of the Whooping Crane*. University of Texas Press: Austin, TX.

Guravich, Dan. 1983. *The Return of the Brown Pelican*. Louisiana State University Press: Baton Rouge, LA.

Mark, R. Stanley Price. 1989. *Animal Re-Introductions: The Arabian Oryx in Oman.* Cambridge University Press: New York, NY.

Mooney, Harold A. and Richard J. Hobbs. 2000. *Invasive Species in a Changing World.* Island Press: Covelo, CA.

Office of Insular Affairs. 1999. *Integrated Pest Management Approaches to Preventing the Dispersal of the Brown Tree Snake and Controlling Snakes in Other Situations.* United States Department of the Interior: Washington, DC.

Simons, Ted, Steve K. Sherrod, Michael W. Collopy, and M. Alan Jenkins. 1997. Restoring the bald eagle. Pp. 229–237 in Peter Kareiva, ed., *Exploring Ecology and Its Applications.* Readings from Scientific American. Sinauer Associates: Sunderland, MA.

Snyder, Noel F.R. 2000. *The California Condor: A Saga of Natural History and Conservation.* Academic Press: London, England.

Chapter 5

Involvement Groups

We can all play our individual part in maintaining healthy ecosystems; however, working in groups can greatly increase our influence and enjoyment. One way to get involved is to visit a computer network such as EcoNet or EnviroNet and share your enthusiasm, opinions, and insights with others.

Starting or Joining an Involvement Group

Many counties in New York have an Environmental Management Council comprised of volunteer citizens. Also, the town Planning Commission or Planning Department has public meetings with opportunities to volunteer and suggest projects. Check with your local and regional government agencies on how to provide input and volunteer assistance in your area. The following website gives detailed environmental profiles of communities, including levels of pollution, extent of wildlife, recycling information, and local gardening:

http://www.formyworld.com
(then type in the zipcode of the area you are considering)

There are also community projects that allow participation. The Audubon Society, for example, surveys where birds are found and how many birds are seen at different times of the year. There are also likely to be local botanical and herpetological clubs active in environmental affairs.

Groups on campus, such as Beta Beta Beta National Biological Honors Society, or a local Science Club may undertake environmentally oriented projects. These projects may include writing a field guide for a local area, cleaning up along beaches and highways, planting trees, building or installing nest boxes, and raising funds for environmental causes.

Public Interest Research Groups (PIRGs) unite students and professional staff (lawyers, researchers, and scientists) to solve problems in their community. There are more than 100 chapters on campuses across the country.

Another national student group is the Student Environmental Action Coalition (SEAC), with over 450 member groups in all 50 states and some foreign countries. Local and national letter-writing campaigns are used to help shape political policy in environmental issues, and the *Network News* newsletter informs members of issues and ways of dealing with them.

If an environmental interest group does not exist on campus, you may want to start one. National Wildlife has sponsored such projects as "Cool It," which seeks to slow global warming. Contacting a regional or national organization such as Greenpeace or the Sierra Club will help in starting a local chapter at your school.

Procedures that you can use to form a group of environmentalists include:

✓ *Enlist:* Talk to friends and friends of friends to determine similar interests and goals

✓ *Grow:* Involve others by setting up a table with posters, literature, and sign-up sheets

✓ *Set a timetable:* Establish a regular time and place, and have interesting speakers and demos

✓ *Join in:* Participate in or establish recycling programs and energy audits

✓ *Teach and promote:* Make announcements of your activities to classes, campus groups, service organizations, places of worship, and newspapers; share your expertise with others; demonstrate peacefully to focus attention on particularly important environmental issues

Useful Addresses for Involvement Groups

The following list includes citations that are either familiar to the authors or sites brought to their attention. Any further suggestions are welcomed.

Beta Beta Beta National Biological Honors Society, P.O. Box 428, Ocean Grove, NJ 07756
732-988-8551
http://tri-beta.org *and* tribeta@bellatlantic.net

EcoNet, 18 DeBoom St., San Francisco, CA 94107
415-442-0220
http://www.igc.org/irg/gateway/ enindex.html *and* econet@igc.apc.org

Greenpeace (this action-oriented group is one of the largest and best-known organizations), 702 H St. NW, Washington, DC 20001
800-326-0959
http://www.greenpeace.org/ *and* inter@ct

National Audubon Society, 700 Broadway, New York, NY 10003
212-979-3000
http://www.audubon.org *and* join@audubon.org

Nature Conservancy, 4245 N. Fairfax Drive, Suite 100, Arlington, VA 22203 (find local chapter)
800-628-6860
http://nature.org *and* comment@tnc.org

Public Interest Research Groups (PIRGs), 218 D St. SE, Washington, DC 20003
202-546-9707

http://www.pirg.org *and*
uspirg@pirg.org

Sierra Club, 408 C St. NE, Washington,
DC 20002 (contact local chapter),
85 Second St., San Francisco, CA 94105
    415-977-5500
    http://www.sierraclub.org *and*
    information@sierraclub.org

Student Environmental Action Coalition
(SEAC), P.O. Box 31909, Philadelphia,
PA 19104
    215-222-4711
    http://www.seac.org/ *and* seac@seac.org

World Environment Organization
HomePage 100 Top Environmental Sites
    http://www.100topenvironment.com/
    Environment/environet/100/66k *and*
    editor@100.com

Compromise and Points of Leverage

It is not likely that every battle fought will
be won. Rather than "cry wolf" at every
proposal, we can save our energies and
resources for items that have been most
highly prioritized. As an example, many
areas in the world have serious ecological
problems. It would be naïve to think that
all problems can be solved within a short
period. The next-best strategy is to identify
areas that are of particular ecological signif-
icance, especially because of biodiversity,
and rally support for these "hot spots."

Arguments for conservation can appeal
to national pride and the promotion of
ecotourism. Every country has something
ecologically unique because no two ecosystems are exactly the same. Each ecosystem
develops at a different rate under different
conditions and has its own unique value
and appeal. Preserving resources is therefore a way of sustaining income, rather
than providing only one payday.

After expressing your concerns and
ideas to your legislator, a follow-up letter is
normal procedure to thank your representative if the response to your letter was positive or to convey your disappointment if
your representative was not persuaded. Also
include suggestions for ways in which
improvement can be made in voting on
environmental issues.

Meeting with your representative can be
especially persuasive. Contact the legislator's
Washington office to determine when the
legislator will be back in the district and call
the local office to schedule an appointment.

Local officials (county commissioners,
city council members, mayors) can also be
approached on issues of environmental
concern. Attendance at public meetings is
a good way to ask questions and propagate
your viewpoints.

Elect officials who will support your
views. Ask friends and family to vote for
environmentally informed and concerned
candidates. A yearly National Environmental Scorecard, published by the League of
Conservation Voters, lists the voting records
of lawmakers on environmental legislation
during the year; a cumulative score is also
given for each of the past three years.

The National Wildlife Federation
(NWF) has an Activist Kit containing booklets on how to influence legislators, information on the effective use of the media,
addresses and directories of all senators and
representatives, and a list of important
committees and subcommittees. The NWF
also has a Resource Conservation Alliance
for members interested in influencing votes

on environmental matters such as wildlife. A semimonthly publication, *Conservation 90*, and *Action Alerts* update methods for influencing the vote of administrators and legislators.

Influencing Legislators

Legislators can be persuaded to support environmental legislation for the protection of species and habitats, provide funds for parks, allot funds for research and education, and address many other areas of ecological concern. A single letter may cause a representative to rethink a position.

Remember to make your letter respectful, factual, to the point, and sincere. As you look at the circled numbers on the sample letter that follows, refer to the numbers on the list below.

① *Do not use form letters or preprinted cards.* Show your concern by writing legibly or typing neatly on your personal letterhead or stationery. Your return address must be included.

② *Address the letter properly*, as shown below. (You can find the name of your representative by calling your public library.)

③ *Focus your letter on one issue.* A bill number (such as H.R. 301) or popular name (for example, Superfund) will help to identify your specific concern and avoid confusion with other bills.

④ *Contact the official just before the vote.* Your points will then be fresh in the legislator's mind. Some conservation groups have hotlines providing updates on what is happening in the legislature.

⑤ *Be precise in your request.* State whether you wish the representative to cosponsor a bill, to vote pro or con against bills, or whether you simply want information on the legislator's position.

⑥ *Be courteous and factual.* Try to provide information or considerations that the representative may not have been informed of, or at least present the information in a novel way that will provoke further thought by the legislator. Do not be threatening.

⑦ *Mention the representative's voting record and public comments on the issue*, if it might help get the legislation passed.

⑧ *Keep a positive tone.* Thank the legislator for hearing your point of view and for any initiative that has been shown in support of your position.

⑨ *Be concise.* Any more than a one-page letter is likely to be overlooked, simply skimmed, or put on permanent hold.

The President The White House 1600 Pennsylvania Ave., NW Washington, DC 20500 202-456-1414 Dear Mr. President,	The Honorable _____ U.S. Senate Washington, DC 20510 202-224-3121 Dear Senator _____ ,	The Honorable _____ U.S. House of Representatives Washington, DC 20515 202-224-3121 Dear Representative _____ ,

Sample Letter to a Legislator

 **From the desk of JOHN Q. PUBLIC** ①

The Honorable Jane Doe
U.S. Senate
Washington, DC 20510

Dear Senator Doe, ②

The U.S. Forest Service has clearcut our public lands and sold timber from our National Forests at below cost for some time. The problems with this practice have become particularly acute in the Tongass Forest of Alaska where the Forest Service spends $40 million a year to promote the clearcutting of some of our oldest and finest trees. In order to restore financial responsibility and to protect a part of the largest remaining temperate rainforest in North America, Congress is considering Bill S. 346, the Tongass Timber Reform Act. ③ This bill is currently in committee and should come up for a vote within the next few weeks. ④

Your cosponsorship and strong support for this bill ⑤ are important for the following reasons. During 1985 and 1986 the Forest Service earned only one cent for every dollar it spent on timber sales in the Tongass. Despite massive public subsidies, timber industry employment in the area has fallen due to increased mechanization. In addition, these trees represent the last remaining large tracts of old-growth forest in the U.S. The House Bill passed last summer (H.R. 987) would protect 1.8 million acres of the Tongass as wilderness. The Senate should do the same, since less than 2 percent of all lands in the U.S. are designated wilderness areas and the amount of land qualified for designation shrinks every year. ⑥

Records show that you did not cosponsor this bill in 1988 or 1989. ⑦ However, your recent rating by the League of Conservative Voters shows that you are increasingly responsive to the environmental concerns of your constituents. ⑧ I hope that you will continue to demonstrate progressive leadership by cosponsoring the Tongass Timber Reform Act in 1990.

Sincerely,

John Q. Public

John Q. Public
Address ⑨

Modified from Causey, 1991

References for Involvement Groups

The following lists include citations that are either familiar to the authors or sites brought to their attention. Any further suggestions are welcomed.

Causey, Ann S. 1991. *Environmental Action Guide: Action for a Sustainable Future*. Benjamin/Cummings Publishing: Redwood City, CA.

Clabaugh, Gary K. and Edward G. Rozycki. 1997. *Analyzing Controversy: An Introductory Guide*. Dushkin/McGraw-Hill: Guilford, CT.

Congressional Record Online via GPO Access (listing bills and their sponsors)
888-293-6498 or 202-512-1262
http://www.access.gpo.gov/ su_docs/aces/aces150.html *and* gpoaccess@gpo.gov

Environmental Defense Toxic Release Inventory Database
http://environmentaldefense.org

Goldfarb, Theodore D. 1997. *Clashing Views in Controversial Environmental Issues*. Dushkin/McGraw-Hill: Guilford, CT.

Hickman, Graham C., ed. 1995. *A Field Guide to Ward Island*. Epsilon Phi Chapter of Beta Beta Beta: Corpus Christi, TX.

Honey, Martha. 1999. *Ecotourism and Sustainable Development: Who Owns Paradise?* Island Press: Covelo, CA.

League of Conservation Voters, 1920 L St. NW, Suite 800, Washington, DC 20036
202-785-8683
http://www.lcv.org *and* lcv.org

Natural Resources Defense Council, 40 West 20th St., New York, NY 10011 (free e-mail service for reports on current issues and bills)
212-727-2700
http://www.nrdcorg *and* nrdcinfo@nrdcorg

National Wildlife Federation, Office of Legislative Affairs, 1400 16thSt. NW, Washington, DC 20036 (the Activist Kit lists congressional committees and personal addresses for correspondence with your congressman/woman on specific issues)

Prendergast, J.R., R.M. Quinn, J.H. Laton, B.C. Eversham, and D.W. Gibbons. 1993. Rare species, the coincidence of diversity hotspots and conservation strategies. *Nature* 365: 335–337.

Chapter 6

References, Directories, Organizations, and Other Resources

Rather than trying to be exhaustive and listing thousands of titles, the following sampling of references is meant to give a general idea of the types and variety of resources available. Scanning and becoming familiar with the reserve shelves and the ecology section of local and college libraries is a good way to learn about resources.

General Ecological References

Allaby, Michael, ed. 2000. *A Dictionary of Ecology*. NetLibrary, Inc.: Boulder, CO.
http://www.netlibrary.com/
summary.asp?ID=12304

Allen, T.F.H. and Thomas W. Hoekstra. 1992. *Toward a Unified Ecology*. Columbia University Press: New York, NY.

Art, Henry W., ed. 1993. *The Dictionary of Ecology and Environmental Science*. H. Holt: New York, NY.

Attenborough, David. 1984. *The Living Planet: A Portrait of the Earth*. Little, Brown: Boston, MA.

Barrows, Edward M. 2001. *Animal Behavior Desk Reference: A Dictionary of Animal*

Behavior, Ecology, and Evolution. CRC Press: Boca Raton, FL.

Calow, Peter, ed. 1998. *The Encyclopedia of Ecology and Environmental Management*. Blackwell Science: Malden, MA.

Calow, Peter, ed. 1999. *Blackwell's Concise Encyclopedia of Environmental Management*. Blackwell Science: Malden, MA.

Carson, Rachel. 1962. *Silent Spring*. Houghton Mifflin: Boston, MA.

Colinvaux, Paul. 1978. *Why Big Fierce Animals Are Rare: An Ecologist's Perspective*. Princeton University Press: Princeton, NJ.

Collin, P.H. 1998. *Dictionary of Ecology and the Environment*. Fitzroy Dearborn Publishers: Chicago, IL.

Farb, P. 1970. *Ecology*. Time-Life Books: New York, NY.

Grzimek, Bernhard, ed. 1976. *Grzimek's Encyclopedia of Ecology*. Van Nostrand: New York, NY.

Hanson, Herbert C. 1962. *Dictionary of Ecology*. Philosophical Library: New York, NY.

Kareiva, Peter, ed. 1997. *Exploring Ecology and Its Applications: Readings from American Scientist*. Sinauer Associates: Sunderland, MA.

Leopold, Aldo. 1949. *A Sand County Almanac, and Sketches Here and There*. Oxford University Press: New York, NY.

Lincoln, R.J., G.A. Boxshall, and P.F. Clark. 1998. *A Dictionary of Ecology, Evolution and Systematics*. Cambridge University Press: New York, NY.

Periodicals

The American Journal of Alternative Agriculture (avoiding pesticides)
Henry A. Wallace Institute for Alternative Agriculture, 38 Winrock Dr., Morrilton, AR 72110
501-727-5435
http://winrock.org/wallacecenter/ajaa.htm *and* hawiaa@access.digex.net

The Amicus Journal (natural resource issues)
Natural Resources Defense Council, Inc., 40 West 20th St., New York, NY 10011

Archives of Environmental Health: An International Journal (general public coverage)
Heldref Publications, 1319 18th St. NW, Washington, DC 20036
202-296-6267
http://www.heldref.org *and* aeh@heldref.org

Audubon (all environmental issues with wonderful photography)
National Audubon Society, 700 Broadway, New York, NY 10003

212-979-3000
http://www.audubon.org *and* join@audubon.org

Bulletin of the Atomic Scientists (covers nuclear issues)
Educational Foundation for Nuclear Science, 6040 S. Kimbark Ave., Chicago, IL 60637
773-702-2555
http://www.thebulletin.org/index.html *and* bulletin@thebulletin.org

Conservation Biology
Blackwell Scientific Publications, Commerce Place, 350 Main St., Malden, MA 02148
888-661-5800
http://conbio.net/scb/journal *and* conbio@u.washington.edu

Conservation Foundation Letter (newsletter)
Conservation Foundation, 1717 Massachusetts Ave. NW, Washington, DC 21136

The Ecologist
MIT Press Journals, 55 Hayward St., Cambridge, MA 02142

The Environmental Magazine (bimonthly publication)
Earth Action Network, Inc., Norwalk, CT

Earth Island Journal
Earth Island Institute, 300 Broadway, Suite 28, San Francisco, CA 94133
415-788-3666 x 123
http://www.earthisland.org/eijournal/journal.cfm *and* earthisland@earthisland.org

Environmental Action (one of the oldest sources)

    6930 Carroll Ave., Suite 600, Takoma Park, MD 20912

    301-891-1100

Environmental Ethics (written for a wide audience with some technical philosophy)

    Center for Environmental Philosophy, P.O. Box 310980, University of North Texas, Denton, TX 76203

    940-565-2727

    http://www.cep.unt.edu/enethics.html *and* cep@unt.edu

The Futurist (readable accounts on societal issues)

    World Future Society, 7910 Woodmont Ave., Suite 450, Bethesda, MD 20814

    800-989-8274 or 301-656-8274

    http://wfs.org/futurist.htm *and* sechard@wfs.org

Green Earth Observer (has good reviews of books and periodicals)

    Green Earth Foundation, P.O. Box 327, El Verano, CA 95433

Greenpeace Magazine (one of the best-known conservation organizations)

    Greenpeace, USA, Inc., 702 H St. NW, Washington, DC 20001

    800-326-0959

    http://www.greenpeace.org/ *and* inter@ct

International Wildlife (photography outstanding)

    National Wildlife Federation, 11100 Wildlife Center Drive, Reston, VA 20190

    800-822-9919

    http://www.nwf.org/

Living Wilderness (issues involving land use)

    The Wilderness Society (preservation of wilderness and wildlife; free *World Alerts*), 1615 M St. NW, Washington, DC 20036

    800-THE-WILD

    http://www.wilderness.org *and* tws@wilderness.org

Journal of Wildlife Management

    Wildlife Society, 5410 Grosvenor Lane, Bethesda, MD 20814

    301-897-9770

    http://www.wildlife.org/ *and* tws@wildlife.org

Mother Earth News (organic farming and articles on environmental friendly lifestyles)

    P.O. Box 56302, Boulder, CO 80322

    800-234-3368

    http://www.motherearthnews.com *and* letters@motherearthnews.com

National Parks and Conservation Magazine (protection of our national parks)

    National Parks & Conservation Association, 1300 19th St. NW, Washington, DC 20036

    http://www.npca.org/magazine/ *and* npca@npca.org

National Wildlife (*Action Alerts* with notable photography)

    National Wildlife Federation, 11100 Wildlife Center Drive, Reston, VA 20190

    800-822-9919

    http://www.nwf.org/

Natural History (field science and conservation coverage)

    American Museum of Natural History, Central Park West at 79th St., New York, NY 10024

    212-760-5100

http://www.amnh.org/naturalhistory/
and nhmag@amnh.org

Nature Conservancy News (focus on natural areas)
Nature Conservancy, 4245 N. Fairfax
Drive, Suite 100, Arlington, VA 22203
800-628-6860
http://nature.org *and*
comment@tnc.org

Not Man Apart (protectionist's view on current events in newspaper style)
Friends of the Earth, 1025 Vermont
Ave., Washington, DC 20005
887-843-8687
http://www.foe.org *and* foe@foe.org

Organic Gardening (techniques and tips)
Organic Gardening, Box 7320, Red Oak,
IA 51591
800-666-2206
http://www.organicgardening.com *and*
organicgardening@rodale.com

Population Bulletin (very readable)
Population Reference Bureau,
1875 Connecticut Ave. NW, Suite 520,
Washington, DC 20009
800-877-9881
http://www.prb.org/pubs/
population_bulletin/ *and*
popref@prb.org

Science News (weekly summaries with environment section)
Science News Service, Inc., 1719 N. St.
NW, Washington, DC 20036
202-785-2255
http://www.sciencenews.org *and*
scinews@sciserv.org

Sierra (writing and photography exceptional)
Sierra Club, 85 Second St., San

Francisco, CA 94105
415-977-5500
http://www.sierraclub.org *and*
information@sierraclub.org

State of the World (environmental trends published annually)
WorldWatch Institute, 1776 Massachusetts Ave. NW, Washington, DC 20036
202-452-1999
http://worldwatch.org *and*
worldwatch@worldwatch.org

The Trumpeter: Journal of Ecosophy (quarterly thought-provoker)
Athabasca University, 1 University Dr.,
Athabasca, Alberta, Canada T9S 3A3
800-780-9041

Utne Reader (not exclusively environmental in topic)
1624 Harmon Place, Minneapolis,
MN 55403
612-338-5048
http://www.utne.com *and*
editor@mail.utne.com

Volunteer Monitor (volunteer water quality monitoring; free newsletter)
River Network, 520 SW 6th Ave., Suite
1130, Portland, OR 97204
503-241-3506
Volmon@rivernetwork.org *and*
http://www.epa.gov/owow/volunteer/
vm_index.html (newsletter)

Wilderness (strong conservation ethic)
The Wilderness Society (preservation of
wilderness and wildlife; free *World Alerts*)
1615 M St. NW, Washington, DC 20036
800-THE-WILD
http://www.wilderness.org *and*
tws@wilderness.org

WorldWatch (each issue looks to change the course of history; one of the best)

WorldWatch Institute, 1776 Massachusetts Ave. NW, Washington, DC 20036

202-452-1999

http://www.worldwatch.org *and* worldwatch@worldwatch.org

Media

There are too many worthwhile films, videos, CDs, and audio recordings to review them all, or even list them here by title. If there are certain films that you consider praiseworthy, please send the title, company, date, length, comments, and your rating to the address listed in the Preface (four stars for the highest rating, three stars for above average, two stars for average, and one star for the lowest rating). Please state whether you consent to having your report back-published with your name or initials.

Bullfrog Films, 372 Dautrich Road, Reading, PA 19606

610-779-8226

http://www.bullfrogfilms.com *and* catalog@bullfrogfilms.com

Carolina Biol. Supply Co., Environmental Science and Ecology, 2700 York Rd., Burlington, NC 27215

800-334-5551

http://www.carolina.com *and* carolina@carolina.com

EnviroVideo, Box 311, Ft. Tilden, NY 11695

800-ECO-TV46

http://home.earthlink.net/~envirovideo *and* envirovideo@earthlink.net

Films for the Humanities and Sciences Catalog, P.O. Box 2053, Princeton, NJ 08543

800-257-5126

http://www.films.com *and* cust.serv@films.com

JLM Visuals, 920 Seventh St., Grafton, WI 53024

262-377-7775

http://www.jlmvisuals.com *and* info@jlmvisuals.com

National Film Board of Canada, 350 Fifth Ave., Suite 4820, New York, NY 10118

212-629-8890

http://www.nfb.ca/ *and* NewYork@nfb.ca

Teacher's Video Company, P.O. Box 4455, Scottsdale, AZ 85261

800-262-8837

http://www.teachersvideo.com *and* customerservice@teachersvideo.com

There is some spectacular footage of habitats and inhabitants in *The Living Planet* series with David Attenborough (BBC/Time-Life Books).

Directories and Agencies

Amazing Environmental Organization Web Directory (largest environmental search engine)

http://www.webdirectory.com

Annual Environmental Sourcebook

Island Press, 58440 Main St., P.O. Box 7, Covelo, CA 95428

800-828-1302

http://www.islandpress.org *and* ipwest@islandpress.org

Bureau of Land Management, U.S. Department of the Interior, 1849 C Street, Room 406-LS, Washington, DC 20240
202-452-5125
http://www.blm.gov/nhp/index.htm
and woinfo@blm.gov

Bureau of Reclamation, Director of Research and Natural Resources, 1849 C St. NW, Washington, DC 20240
202-513-0682
http://www.usbr.gov/main/

Congressional Research Service, National Council for Science and the Environment, 1725 K Street NW, Suite 212, Washington, DC 20006
202-530-5810

Conservation Directory (organizations, agencies, and officials updated annually)
National Wildlife Federation, 11100 Wildlife Center Drive, Reston, VA 20190
800-822-9919
http://www.nwf.org/printandfilm/publications/consdir/

Directory of Environmental Organizations (approximately 6,600 names and addresses)
Educational Communications, P.O. Box 351419, Los Angeles, CA 90035
310-559-9160
http://www.ecoprojects.org/ecoprojects *and* ENCP@aol.com

Environmental Protection Agency, 1200 Pennsylvania Ave. NW, Washington, DC 20460
202-382-2090
http://www.epa.gov *and* public-access@epa.gov

Environmental Research Information System (UFIS), GSF (Forschungszentrum fuer Umwelt und Gesundheit), Neuherberg, Postfach 1129, Germany
http://www.gsf.de/UNEP/contents.html *and* ighe@gsf.org

FAO (Food and Agriculture Organization of the United Nations)
Viale delle Terme di Caracalla, 00100 Rome, Italy
39-0657051
http://www.fao.org/INSIDE/faohqe.htm *and* webmaster@fao.org

Fish and Wildlife Service, Department of the Interior, 18th and C Sts. NW, Washington, DC 20240
http://www.fws.gov *and* contact@fws.gov

GETF (Global Environmental and Technology Foundation), 7010 Little River Turnpike, Suite 460, Annandale, VA 22003
703-750-6401
http://www.getf.org *and* getf@getf.org

International Union for Conservation of Nature and Natural Resources (IUCN), World Conservation Union, Rue Mauverney 28, 1196 Gland, Switzerland
41-22-999-0001
http://www.iucn.org/2000/about/content/people/index.html *and* ursula.hiltbrunner@iucn.org

Internet Reference Desk, Environment and Natural Resource Protection
http://www.familyhaven.com/Internetreferencedesk/enrpage.html

National Center for Atmospheric Research, 1850 Table Mesa Drive, Boulder, CO 80305
303-497-1000

National Council for Science for the Environment, 1725 K St. NW, Suite 212, Washington, DC 20006
    202-530-5810
    http://www.cnie.org/nle/clim-7/ index.html *and* info@NCSEonline.org

National Park Service, Department of the Interior, 1849 C St. NW, Washington, DC 20240
    202-208-6843
    http://www.nps.gov/

NOAA (National Oceanic and Atmospheric Administration), 14th St. and Constitution Ave. NW, Room 6013, Washington, DC 20230
    202-482-6090
    http://www.noaa.gov *and* answers@noaa.gov

Office of Ocean and Coastal Resource Management, 1825 Connecticut Ave., Suite 700, Washington, DC 20235
    301-713-3155
    http://www.ocrm.nos.noaa.gov/ *and* Elaine.Vaudreuil@noaa.gov

USGS (United States Geological Survey) Monitoring Program, Patuxent Wildlife Research Center, 12100 Beech Forest Road, Suite 4039, Laurel, MD 20708
    301-497-5500
    http://www.pwrc.usgs.gov/ *and* director@patuxent.usgs.gov

United States and the Global Environment: A Guide to American Organizations Concerned with International Environmental Issues (summarizes private and government groups). 1983. Thaddeus C. Trzyna, ed. California Institute of Public Affairs: Claremont, CA.

United States Department of Agriculture Forestry Service, P.O. Box 96090, Washington, DC 20090
    202-205-1680
    http://www.fs.fed.us/ *and* wo_fs-contact@fs.fed.us

World Directory of Environmental Organizations, California Institute of Public Affairs, P.O. Box 189040, Sacramento, CA 95818
    916-442-2462
    cipa@cipahq.org *and* info@cipahq.com

Ecologically Concerned Organizations

Air Pollution Control Association (international)
    P.O. Box 2861, Pittsburgh, PA 15230
    412-232-3444
    http://www.nsie.org/lib5.htm

American Cetacean Society (whales, dolphins, porpoises)
    P.O. Box 1391, San Pedro, CA 90733
    310-548-6279
    http://www.acsonline.org/ *and* acs@pobox.com

American Conservation Association
    1350 New York Ave. NW, Suite 300, Washington, DC 20005
    202-624-9365
    74111.3156@compuserve.com

American Farmland Trust (preserving our agricultural land base and sustainability)
    1200 18th St. NW, Suite 800, Washington, DC 20036
    202-331-7300
    http://www.farmland.org *and* info@farmland.org/

American Forestry Association (plus water and air issues; *Global Releaf*)
> P.O. Box 2000, Washington, DC 20013
> 202-955-4500
> http://www.americanforests.org/ *and* info@amfor.org

American Rivers (also includes landscapes)
> 1025 Vermont Ave. NW, Suite 720, Washington, DC 20005
> 202-347-7550
> http://www.amrivers.org/ *and* amrivers@amrivers.org

Center for Marine Conservation
> 1725 De Sales St. NW, Suite 600, Washington, DC 20036
> 202-429-5609
> http://www.cmc-ocean.org *and* cmc@dccmc.org

Center for Science in the Public Interest (focus on sustainability)
> 1875 Connecticut Ave. NW, Suite 300, Washington, DC 20009
> 202-332-9110
> http://www.cspinct.org/ *and* cspi@cspinet.org

Citizens' Clearinghouse for Hazardous Wastes (for towns in pollution legalities)
> P.O. Box 6806, Falls Church, VA 20040
> 703-276-2249

Clean Water Action Project (a political action–oriented group)
> 4455 Connecticut Ave. NW, Suite A300, Washington, DC 20008
> 202-895-0420
> http://www.cleanwateraction.org *and* CWA@CleanWaterAction.org

Concern, Inc. (makes available information regarding the environment)
> P.O. Box 1133, Washington, DC 20013

> 800-336-4797
> http://www.health.gov/NHICScripts/ Entry.cfm?HRCode=HR0058 *and* concern@igc.org

Conservation International (focuses on protection of tropical ecosystems)
> 1919 M St. NW, Suite 600, Washington, DC 20036
> 800-406-2306 or 202-912-1000
> http://www.conservation.org

Conservation Law Foundation (legalities in support of U.S. natural resources)
> 62 Summer St., Boston, MA 02110
> 617-350-0990
> http://www.clf.org *and* ihart@clf.org

The Cousteau Society (focuses on world-wide marine resources)
> 870 Greenbrier Circle, Suite 402, Chesapeake, VA 23320
> 800-411-4395
> http://www.cousteausociety.org *and* cousteau@cousteausociety.org

Defenders of Wildlife (protection of species and habitats)
> 1101 14th St. NW, #1400, Washington, DC 20005
> 202-682-9400
> http://www.defenders.org *and* info@defenders.org

Earth First! (has used controversial means to protect the environment)
> P.O. Box 3023, Tucson, AZ 85702
> 520-620-6900
> http://www.earthfirstjournal.org *and* collective@earthfirstjournal.org

Earth Island Institute (international politics and the environment)
> 300 Broadway, Suite 20, San Francisco, CA 94133

415-788-3666
http://www.earthisland.org *and*
earthisland@earthisland.org

Earthsystems
508 Dale Ave., Charlottesville, VA 22903
804-293-2022
http://www.earthsystems.org/
index.html *and*
www@earthsystems.org

Environmental Action, Inc. (political and social change regarding environmental issues)
1525 New Hampshire Ave. NW, Washington, DC 20036
202-745-4870

Environmental Defense Fund (economists, lawyers, and scientists for the public good)
257 Park Ave. South, New York, NY 10010
212-505-2100
http://www.edf.org

Environmental Management (journal)
Springer-Verlag New York, 175 Fifth Ave., New York, NY 10010
800-SPRINGER
http://www.springer-ny.com *and*
service@springer-ny.com

Friends of the Earth/Environmental Policy Institute/Oceanic Society (large and effective)
1025 Vermont Ave. NW, Washington, DC 20005
877-843-8687 or 202-783-7400
http://www.foe.org *and* foe@foe.org

The Greens (Green Party USA is a political conservation group)
P.O. Box 100, Blodgett Mills, NY 13738
866-GREENS2
http://www.greenparty.org *and*
gpusa@greens.org

Izaak Walton League (supports conservation education)
IWLA Membership Department, 707 Conservation Lane, Gaithersburg, MD 20878
800-IKE-LINE
http://www.iwla.org *and* general@iwla.org

League of Conservation Voters (supports election of pro-environmental candidates)
1920 L St. NW, Suite 800, Washington, DC 20036
202-785-8683
http://www.lcv.org

National Audubon Society (research, education, advocacy of conservation/preservation)
700 Broadway, New York, NY 10003
212-979-3000
http://www.audubon.org *and*
join@audubon.org

National Parks and Conservation Association (protection and betterment of our parks)
1300 19th St. NW, Suite 399, Washington, DC 20036
202-944-8530
http://www.npca.org *and*
npca@npca.org

National Wildlife Federation (over 6 million members and supporters)
11100 Wildlife Center Drive, Reston, VA 20190
800-822-9919
http://www.nwf.org/

Natural Resources Defense Council (scientific research and lawsuits against polluters)
40 West 20th St., New York, NY 10011
212-727-2700
http://www.nrdcorg *and*
nrdcinfo@nrdcorg

Nature Conservancy
   4245 N. Fairfax Drive, Suite 100,
   Arlington, VA 22203
   800-628-6860
   http://nature.org *and*
   comment@tnc.org

Population-Environment Balance (Popline
Newsletter)
   2000 P St. NW, Suite 600, Washington,
   DC 20036
   202-955-5700
   http://www.balance.org *and*
   uspop@us.net

The Population Institute (Popline Newsletter)
   107 Second St. NE, Washington,
   DC 20002
   202-544-3300 ext. 109
   http://www.populationinstitute.org *and*
   etars@populationinstitute.org

Population Reference Bureau (good source
for demographic data on population
growth)
   1875 Connecticut Ave. NW, Suite 520,
   Washington, DC 20009
   202-483-1100
   http://www.prb.org *and* popref@prb.org

RAN (Rainforest Action Network)
   221 Pine St., Suite 500, San Francisco,
   CA 94104
   415-398-4404
   http://www.ran.org/ran/ *and*
   rainforest@ran.org

Sierra Club (a centenarian, this well-
respected organization focuses on land use)
   85 Second St., 2nd Floor, San Francisco,
   CA 94105
   415-977-5500

http://www.sierraclub.org *and*
information@sierraclub.org

Smithsonian Institution (education and
research programs over a wide range)
   1000 Jefferson Drive SW, Washington,
   DC 20560
   202-357-2700
   http://www.si.edu *and* info@si.edu

Union of Concerned Scientists (citizens and
scientists noting the impact of technology)
   2 Brattle Street, Cambridge, MA 02238
   617-547-5552
   http://www.ucsusa.org/energy/
   energy-home.html *and* ucs@ucsusa.org

United Nations Environment Program (protection programs for the environment)
   One United Nations Plaza, New York,
   NY 10017
   212-906-5000
   http://www.unep.org

The Wilderness Society (preservation of
wilderness and wildlife; free *World Alerts*)
   1615 M St. NW, Washington, DC 20036
   800-THE-WILD
   http://www.wilderness.org *and*
   tws@wilderness.org

Wildlife Habitat Enhancement Council
(making corporate-held lands better for
wildlife)
   1010 Wayne Ave., Silver Spring,
   MD 20910
   301-588-8994
   http://www.webdirectory.com/Wildlife

Wildlife Management Institute (education
and research oriented)
   1101 14th St. NW, Suite 801, Washington, DC 20005

202-371-1808
http://www.wildlifemgt.org *and*
wmiyf@ionet.net
http://enviroyellowpages.com/
maryland/maryland24.htm

WorldWatch (research/publication series
on social, economic, and environmental
problems)
1776 Massachusetts Ave. NW,
Washington, DC 20036
202-452-1999
http://www.worldwatch.org *and*
worldwatch@worldwatch.org

World Wildlife Fund (protection of wildlife and habitats, focusing on third world
countries)
1250 24th St. NW, P.O. Box 97180,
Washington, DC 20090
800-CALL-WWF
http://www.worldwildlife.org

(ZPG) Zero Population Growth,
1400 16th St. NW, Suite 320, Washington,
DC 20036
202-332-2200
http://zpg.org *and* info@zpg.org

General Field Equipment and Laboratory Supplies

Advanced Telemetry Systems, 470 1st Ave. N,
Box 398, Isanti, MN 55040
763-444-9267
http://www.atstrack.com/Resources/
Main.html *and* sales@atstrack.com

Avinet (mist nets and banding supplies)
P.O. Box 1103, Dryden, NY 13053
888-284-6387

http://www.avinet.com *and*
avinet@lightlink.com

Ben Meadows Company, P.O. Box 5277,
Janesville, WI 53547 (sampling equipment
and instrumentation)
800-241-6401
http://www.benmeadows.com *and*
mail@benmeadows.com

Carolina Biol. Supply Co., Environmental
Science and Ecology, 2700 York Rd.,
Burlington, NC 27215
800-334-5551
http://www.carolina.com *and*
carolina@carolina.com

Christensen Designs, 535 W. Yosemite
Ave., Manteca, CA 95337 (video
probes)
209-239-7460
http://www.PeeperPeople.com *and*
Ann@PeeperPeople.com

Cole-Parmer Instrument Company,
625 East Bunker Court, Vernon Hills,
IL 60061
800-323-4340
http://www.coleparmer.com/ *and*
info@coleparmer.com

Fischer Chemicals, 7360 Milnor St.,
Philadelphia, PA 19136
888-243-6425
http://www.chemical.net/home *and*
cservice@chemical.net

Fisher Science Education, 485 South
Frontage Road, Burr Ridge, IL 60521
800-955-0740
http://www.fisheredu.com *and*
info@fisheredu.com

Forestry Suppliers, Inc., Enviromental.
Products Catalog, 205 West Rankin St.,
P.O. Box 8397, Jackson, MS 39284
     800-647-5368
     http://www.forestry-suppliers.com *and*
     fsi@forestry-suppliers.com

Frey Scientific, 100 Paragon Parkway,
Mansfield, OH 44903
     1-800-225-3739
     http://www.freyscientific.com *and*
     catalog@freyscientific.com

H.B. Sherman Traps, 3731 Peddle Drive,
Tallahassee, FL 32303
     850-575-8727
     becca@electro-net.com

Sargent-Welsh Scientific Co., P.O. Box 5229,
Buffalo Grove, IL 60089
     800-727-4368
     http://www.sargentwelch.com/index.html
     and sarwel@sargentwelsh.com

Team Labs Probware, 6859 N. Foothills
Hwy., Bldg. D200, Boulder, CO 80302
     800-775-4357
     http://www.teamlabs.com *and*
     help@teamlabs.com

Trailmaster Infrared Trail Monitors, Good-
son & Associates, 10614 Widmer, Lenexa,
KS 66215
     913-345-8555
     http://www.trailmaster.com

Vernier Software and Technology,
13979 SW Millikan Way, Beaverton,
OR 97005 (Environmental recording
instrumentation and laboratory instruction
packages)
     503-277-2299
     http://www.vernier.com

VWR Scientific Products
     800-932-5000
     http://www.vwrsp.com/catalog/ *and*
     solutions@vwrsp.com

Ward's Natural Science Estab., 5100
Henrietta Rd., Box 92912, Rochester,
NY 14692
     USA: 800-962-2660
     Canada: 800-387-7822
     International: 716-334-6174
     http://www.wardsci.com/

Woodstream Corporation, Front and Locust
Sts., Lititz, PA 17543 (Havahart and Victor
Traps)
     800-800-1819 or 717-626-2125
     http://www.havahart.com *and*
     havahartmail@woodstream.com
     http://www.victorpest.com *and*
     consumercare@woodstream.com

Chapter 7

Future Study

Contributions to our understanding of ecosystems will help in avoiding some of the problems ahead and assist in providing answers for some of the ecological difficulties our planet is experiencing in ever-increasing force. Refer to Chapter 1 for further education and employment opportunities.

Ecology Field and Laboratory Exercises and Tools

Bart, Jonathon, M.A. Fligner and W.J. Notz. 1999. *Sampling and Statistical Methods for Behavioral Ecologists*. Cambridge University Press: New York, NY.

Bennett, Donald P. and David A. Humpharies. 1974. *Introduction to Field Biology*. Edward Arnold: London, England.

Brewer, Richard and Margaret McCann. 1982. *Laboratory and Field Manual of Ecology*. Saunders College Publishing: Fort Worth, TX.

Brower, J.E., J.H. Zar and C.N. von Ende. 1998. *Field and Laboratory Methods for General Ecology*. WCB/McGraw-Hill: Boston, MA. (includes "Ecological Quantitative Analysis" software)

Colwell, Robert K. 1997. *Biota: The Biodiversity Database Manager*. Sinauer Associates: Philadelphia, PA. (a graphical interface to a fully relational database)

Cox, George W. 1996. *Laboratory Manual of General Ecology*. McGraw-Hill Higher Education: Boston, MA.

Dale, V.H. and M.R. English. 1998. *Tools to Aid Environmental Decision Making*. Springer-Verlag: New York, NY.

Eckblad, J.W. 1978. *Laboratory Manual of Aquatic Biology*. Wm. C. Brown: Dubuque, IA.

Enger, E.D. and B.F. Smith. 1997. *Field and Laboratory Activities in Environmental Science*. Wm. C. Brown Publishers: Dubuque, IA.

Gilligan, M.R., T. Kozel and J.P. Richardson. 1991. *Environmental Science Laboratory*. Halfmoon Publishing: Savannah, GA.

Hairston, Nelson G. 1997. *Ecological Experiments: Purpose, Design and Execution*. Cambridge University Press: New York, NY.

Hauer, F. Richard and Gary A. Lamberti, eds. 1996. *Methods in Stream Ecology*. Academic Press: San Diego, CA.

Knudsen, J.W. 1978. *Collecting and Preserving Plants and Animals*. Harper & Row: New York, NY.

Kolbe, C.M. and M.W. Luedke. 1993. *A Guide to Freshwater Ecology*. Texas Natural Resource Conservation Commission: Austin, TX.

Krebs, Charles J. 2001. *Ecological Methodology*. Benjamin/Cummings: Menlo Park, CA.

Krumhardt, B.A. and D.M. Wirth. 1999. *Experiences in Environmental Science*. Bellwether-Cross Publishing: East Dubuque, IL.

Lewis, T. and L.R. Taylor. 1967. *Introduction to Experimental Ecology*. Academic Press: London, England.

Meir, Eli. 1996. *EcoBeaker 1.0: An Ecological Simulation Program*. Sinauer Associates: Philadelphia, PA.

Scheiner, S.M. and J. Gurevitch. 2000. *Design and Analysis of Ecological Experiments*. Oxford University Press: New York, NY.

Smith, Robert Leo and Thomas M. Smith. 2001. *Ecology and Field Biology*. Benjamin/Cummings Publishing: San Francisco, CA. (see appendices: A, Sampling Plant and Animal Populations; B, Sampling Plant and Animal Populations)

Southwood, T.R.E. 1971. *Ecological Methods*. Chapman & Hall: London, England.

Sutherland, William J. 1996. *Ecological Census Techniques: A Handbook*. Cambridge University Press: New York, NY.

Tomera, A.N. (Revised by J. Beller.) 1989. *Understanding Basic Ecological Concepts*. J. Weston Walch, Publisher: Portland, ME.

Wolf, R.J., C.B. DeWitt, K. Jankowski and G. Van Dyke. 1993. *Environmental Science in Action*. Saunders College Publishing: Fort Worth, TX.

Wratten, Stephen D. and Gary L.A. Fry. 1980. *Field and Laboratory Exercises in Ecology*. Edward Arnold: London, England.

Researching the Literature

This guide has provided many listings of sources and references that can be used to guide your ecological activities. To keep this information up to date and to expand your areas of interest, further literature sources will be required. Be sure to appreciate the difference between **primary literature** (the original publication) and **secondary literature** (publications that report information found in other sources). Just as information can be distorted in the manner of a rumor being passed on from person to person, it is always advisable to check the original source for the information you require. Science is based on factual and accurate reporting.

   One way to begin a literature search is to browse the bookshelves of your library in the ecology section for general references. These sources give an overview of the topic, and the Literature Cited section at the back of the book or after each chapter will lead to other references of interest; these references will also have references, which will, in turn, also have references. Also, there are scientific abstracts, indices, and various compendiums in the reference section of the library.

The Internet is a quick and convenient way of accessing current information, but you should not overlook hard copy sources. Also, take care to evaluate the websites providing the information. Remember that there are no qualifications to posting a website. Anyone can post any information they choose without any validation. Some sites that provide guidance in site evaluation include:

http://www.library.ucla.edu/libraries/college/
　　instruct/web/critical.htm

http://www.library.cornell.edu/okuref/research/
　　webeval.html

If your library supports the ISI Web of Science database, this is a useful place to begin expanding your literature base. More information is available from:

ISI Web of Science Database,
3501 Market St.,
Philadelphia, PA 19104
　　215-386-0100
　　http://www.isinet.com/demos/
　　webofscience

The specific means for tracking down references available at your local library may vary. Check with your local librarians who are there to help you.

Textbooks in Ecology and Environmental Science

Because of the concise nature of this guide, it will be useful to refer to general references for terminology, further elaboration of concepts, additional principles, and more extensive bibliographic coverage. The year 1995 was subjectively chosen for convenience to separate older works from more recent texts.

General Ecology Textbooks

Pre-1995

Allee, W.C., A.E. Emerson, O. Park, T. Park and K.P. Schmidt. 1949. *Principles of Animal Ecology*. W.B. Saunders Co.: Philadelphia, PA.

Brewer, R. 1994. *The Science of Ecology*. Saunders College Publishing: Fort Worth, TX.

Colinvaux, Paul. 1993. *Ecology 2*. John Wiley & Sons: New York, NY.

Clark, G.L. 1965. *Elements of Ecology*. John Wiley & Sons: New York, NY.

Elton, C.S. 1927. *Animal Ecology*. Macmillan Company: New York, NY.

Krebs, Charles J. 1994. *Ecology: The Experimental Analysis of Distribution and Abundance*. HarperCollins College Publ.: New York, NY.

Lederer, Roger J. 1984. *Ecology and Field Biology*. Benjamin/Cummings: San Francisco, CA.

Odum, Eugene. 1963. *Ecology*. Holt, Rinehart & Winston: New York, NY.

Odum, Eugene P. 1983. *Basic Ecology*. Saunders College Publishing: Philadelphia, PA.

Odum, Eugene P. 1993. *Ecology and Our Endangered Life-Support Systems*. Sinauer Associates: Philadelphia, PA. (later published in 1997 under the title *Ecology*)

Pearse, A.S. 1939. *Animal Ecology*. McGraw-Hill: New York, NY.

Real, Leslie A. and James H. Brown. 1991. *Foundations of Ecology: Classic Papers with Commentaries*. University of Chicago Press: Chicago, IL.

Woodbury, Angus M. 1954. *Principles of General Ecology*. Blakiston: New York, NY.

Worster, Donald. 1994. *Nature's Economy: A History of Ecological Ideas*. Cambridge University Press: New York, NY.

Post-1994

Begon, M., J.L. Harper, and C.R. Townsend. 1996. *Ecology: Individuals, Populations, Communities*. Blackwell Scientific Publishers: Boston, MA.

Bush, Mark. 2000. *Ecology of a Changing Planet*. Prentice-Hall: Upper Saddle River, NJ.

Dodson, S.I., T.F.A. Allen, S.R. Carpenter, A.R. Ives, R.L. Jeanne, J.F. Kitchell, N.E. Langston and M.G. Turner. 1998. *Ecology*. Oxford University Press: New York, NY.

Kormondy, Edward J. 1996. *Concepts of Ecology*. W.B. Saunders Co.: Philadelphia, PA.

Krohne, David. 2001. *General Ecology*. Brooks/Cole Publishing: Belmont, CA.

MacKenzie, A. and A.J. Ball. 1998. *Instant Notes in Ecology*. Springer-Verlag: New York, NY.

Molles, Manuel. 2001. *Ecology: Concepts and Applications*. WCB/McGraw-Hill: Dubuque, IA.

Odum, Eugene P. 1997. *Ecology: A Bridge Between Science and Society*. Sinauer Associates: Sunderland, MA.

Ricklefs, Robert E. 2000. *The Economy of Nature*. W.H. Freeman: New York, NY.

Ricklefs, Robert and Gary Miller. 2000. *Ecology*. W.H. Freeman: New York, NY.

Smith, Robert L. and Thomas M. Smith. 2001. *Elements of Ecology*. Benjamin/Cummings: San Francisco, CA.

Smith, Robert Leo and Thomas M. Smith. 2001. *Ecology and Field Biology*. Benjamin/Cummings: San Francisco, CA.

Stiling, Peter D. 2002. *Ecology: Theories and Applications*. Prentice-Hall: Upper Saddle River, NJ.

Townsend, Colin R., John L. Harper and Michael Begon. 2000. *Essentials of Ecology*. Blackwell Science: Malden, MA.

Ecology Specialty Area Textbooks

Applied Ecology

De Santo, Robert S. 1978. *Concepts of Applied Ecology*. Springer-Verlag: New York, NY.

Newman, Edward I. 1993. *Applied Ecology*. Blackwell Scientific Publishers: London, England.

Behavioral Ecology

Klopfer, P.H. 1973. *Behavioral Aspects of Ecology*. Prentice-Hall: Englewood Cliffs, NJ.

Krebs, J.R. and N.B. Davies. 1993. *An Introduction to Behavioral Ecology*. Blackwell Scientific Publications: London, England.

Evolutionary Ecology

Bulmer, Michael. 1994. *Theoretical Evolutionary Ecology*. Sinauer Associates: Philadelphia, PA.

Pianka, Eric R. 2000. *Evolutionary Ecology*. Benjamin/Cummings: San Francisco, CA.

Landscape Ecology

Naveh, Z. and A.S. Lieberman. 1994. *Landscape Ecology. Theory and Application*. Springer-Verlag: New York, NY.

Turner, Monica Goigel. 2001. *Landscape Ecology in Theory and Practice: Pattern and Process*. Springer: New York, NY.

Limnoecology

Campert, Winfried. 1997. *Limnoecology: The Ecology of Lakes and Streams*. Oxford University Press: New York, NY.

Matthews, William J. 1998. *Patterns in Freshwater Fish Ecology*. Chapman & Hall: New York, NY.

Physiological Ecology

Calow, Peter, ed. 1987. *Evolutionary Physiological Ecology*. Cambridge University Press: New York, NY.

Schmidt-Nielson, Knut. 1975. *Animal Physiology: Adaptation and Environment*. Cambridge University Press: New York, NY.

Townsend, Colin R. and Peter Calow. 1981. *Physiological Ecology: An Evolutionary Approach to Resource Use*. Blackwell Scientific Publications: Oxford, England.

Statistical Ecology

Gauch, Hugh G. 1982. *Multivariate Analysis in Community Ecology*. Cambridge University Press: New York, NY.

Gotelli, Nicholas. 1996. *Null Models in Ecology*. Smithsonian Institution Press: Washington, DC.

Environmental Science Textbooks

The year 1996 was subjectively chosen for convenience to separate older works from more recent texts. Many of these texts have been frequently updated, so that the list of "older" texts is quite small.

Pre-1996

Arms, Karen. 1994. *Environmental Science*. Saunders College Publishing: Fort Worth, TX.

Chiras, Daniel D. 1988. *Environmental Science: A Framework for Decision Making*. Benjamin/Cummings: Menlo Park, CA.

Freedman, Bill. 1995. *Environmental Ecology: The Impacts of Pollution and Other Stresses on Ecosystem Structure and Function*. Academic Press: New York, NY.

Rockett, C.L. and K.J. Van Dellen. 1993. *Living in the Environment, Environmental Science, and Sustaining the Earth*. Wadsworth Publishing Company: Belmont, CA.

Wager, Richard H. 1974. *Environment and Man*. W.W. Norton & Co.: New York, NY.

Post-1995

Botkin, D.B. and E.A. Keller. 2000. *Environmental Science: Earth as a Living Planet.* John Wiley & Sons: New York, NY.

Chiras, Daniel D. 1998. *Environmental Science: A Systems Approach to Sustainable Development.* Wadsworth: Belmont, CA.

Chiras, Daniel D. 2001. *Environmental Science: Creating a Sustainable Future.* Jones & Bartlett: Sudbury, MA.

Chiras, D.D. and J.P. Reganold. 2002. *Natural Resource Conservation: Management for a Sustainable Future.* Prentice-Hall: Upper Saddle River, NJ.

Cunningham, W.P. and B.W. Saigo. 2001. *Environmental Science: A Global Concern.* McGraw-Hill: Boston, MA.

Enger, E.D. and B.F. Smith. 2002. *Environmental Science: A Study of Interrelationships.* McGraw-Hill: Dubuque, IA.

Goudie, Andrew. 2000. *The Human Impact on the Natural Environment.* MIT Press: Cambridge, MA.

Harte, John. 2001. *Consider a Spherical Cow: A Course in Environmental Problem Solving.* University Science Books: Sausalito, CA.

Kaufman, Donald G. and Celia M. Franz. 2000. *Biosphere 2000: Protecting Our Global Environment.* Kendall Hunt Publishing: Dubuque, IA.

La Bonde Hanks, Sharon. 1996. *Ecology and the Biosphere: Principles and Problems.* St. Lucie Press: Delray Beach, FL. (for "non-scientific" students)

Mayer, J. Richard. 2001. *Connections in Environmental Science: A Case Study Approach.* McGraw-Hill: Boston, MA.

McKinney, Michael L. and Robert M. Schoch. 1998. *Environmental Science: Systems and Solutions.* Jones & Bartlett Publishers: Sudbury, MA.

Miller, G.T. 2002. *Living in the Environment: Principles, Connections, and Solutions.* Brooks/Cole: Pacific Grove, CA.

Miller, G. Tyler. 2001. *Environmental Science.* Brooks/Cole: Pacific Grove, CA.

Nadakavukaren, Anne. 2000. *Our Global Environment: A Health Perspective.* Waveland Press: Prospect Heights, IL.

Nebel, Bernard J. and Richard T. Wright. 2000. *Environmental Science: The Way the World Works.* Prentice-Hall: Upper Saddle River, NJ.

Raven, P.H., L.R. Berg and G.B. Johnson. 2000. *Environment.* Harcourt College Publishers: Philadelphia, PA.

Soltzberg, Leonard J. 2001. *The Dynamic Environment.* University Science Books: Sausalito, CA. (computer models given to Harte, 2001)

Southwick, Charles H. 1996. *Global Ecology in Human Perspective.* Oxford University Press: New York, NY.

Turco, Richard P. 1997. *Earth Under Siege: From Air Pollution to Global Change.* Oxford University Press: New York, NY. (for nonscience majors)

Turk, Jonathan. 1998. *Introduction to Environmental Studies.* Saunders College Publishing: Philadelphia, PA.

Wright, Richard T. 2002. *Environmental Science: Towards a Sustainable Future.* Prentice-Hall: Upper Saddle River, NJ.

Index of Organizations